"This volume provides many insightful ideas to address the multiple philosophical debates sparked by the topic of plant cognition. Highly recommended to those who are willing to get into a deep discussion concerning the possibility of non-animal minds."

Miguel Segundo-Ortin, *Utrecht University, The Netherlands*

Philosophy of Plant Cognition

This volume features new research about the philosophy of plant intelligence and plant cognition, one of the most intriguing and complex current debates at the intersection of biology, cognitive science and philosophy.

The debate about plant cognition is marked by deep disagreements. Some theorists are confident that the empirical evidence supports the ascription of cognitive capacities to plants. Others hold that such claims are overblown, and defend more traditional, non-cognitive accounts of plant behavior. Still others seek to formulate intermediate positions. This volume brings together leading researchers from across this theoretical spectrum to tackle the foundational questions that are at issue in the debate about plant cognition. The contributions focus on the philosophical questions raised by recent discoveries and controversies in the empirical sciences, such as: Can plants be said to have genuine cognitive abilities? Can they be characterized as representing or perceiving their environment, as pursuing goals, or even as having some form of conscious experience? Which data could provide evidence for such characterizations? And what are possible implications of these issues for general questions about the nature of cognition, representation, perception, and consciousness?

Philosophy of Plant Cognition will be of interest to scholars and students working in philosophy of mind, philosophy of biology, cognitive science, and plant biology.

Gabriele Ferretti is Assistant Professor of Philosophy at the University of Bergamo (Italy). He is the author of many monographs and journal articles in philosophy of mind and of neuroscience.

Peter Schulte is Associate Professor of Philosophy at Umeå University (Sweden). He is the author of *Mental Content* (2023) and numerous articles on the philosophy of mind and metaphysics.

Markus Wild is Professor of Philosophy at the University of Basel (Switzerland). He mainly works on animal minds and animal ethics.

Routledge Studies in Contemporary Philosophy

For more information about this series, please visit: www.routledge.com/Routledge-Studies
-in-Contemporary-Philosophy/book-series/SE0720

Philosophy of Plant Cognition

Interdisciplinary Perspectives

Edited by Gabriele Ferretti,
Peter Schulte, and Markus Wild

Routledge
Taylor & Francis Group

NEW YORK AND LONDON

First published 2025
by Routledge
605 Third Avenue, New York, NY 10158

and by Routledge
4 Park Square, Milton Park, Abingdon, Oxon, OX14 4RN

Routledge is an imprint of the Taylor & Francis Group, an informa business

© 2025 selection and editorial matter, Gabriele Ferretti, Peter
Schulte, and Markus Wild; individual chapters, the contributors

The right of Gabriele Ferretti, Peter Schulte, and Markus Wild to be
identified as the authors of the editorial material, and of the authors
for their individual chapters, has been asserted in accordance with
sections 77 and 78 of the Copyright, Designs and Patents Act 1988.

ISBN: 978-1-032-49351-0 (hbk)
ISBN: 978-1-032-49352-7 (pbk)
ISBN: 978-1-003-39337-5 (ebk)

DOI: 10.4324/9781003393375

Typeset in Sabon
by Deanta Global Publishing Services, Chennai, India

Contents

Contributors

Marc Artiga is Senior Lecturer of Philosophy at the University of València (Spain). He mainly works on philosophy of biology, philosophy of mind and cognitive science.

Deborah J. Brown is Professor of Philosophy at the University of Queensland (Australia) and Director of the University of Queensland Critical Thinking Project. She is a specialist in the history of philosophy, philosophy of mind and critical thinking and education.

Gabriele Ferretti is Assistant Professor of Philosophy at the University of Bergamo (Italy). He is the author of many monographs and journal articles in philosophy of mind and of neuroscience.

Carrie Figdor is Professor of Philosophy at the University of Iowa (USA). She is the author of *Pieces of Mind* (2018) and works in philosophy of comparative psychology, philosophy of science, philosophy of neuroscience, cognitive science and metaphysics.

P. Adrian Frazier is a doctoral candidate in Ecological Psychology at the University of Connecticut and a member of the Center for the Ecological Study of Perception and Action. He authored several papers, including 'On the Possibility of Plant Consciousness: A View from Ecointeractivism'.

Todd Ganson is Professor of Philosophy at Oberlin College (USA). He works on questions about sensory systems at the intersection of philosophy, biology and psychology.

Brian Key is Professor of Developmental Neurobiology and Head of the Neurophilosophy Lab at the University of Queensland (Australia). He explores conceptual explanations for the neural basis of subjective experience.

Jonny Lee is a teacher of Theology and Philosophy at Christ's Hospital, England. He is the author of numerous articles in philosophy of mind and cognitive science.

Mohan Matthen is Emeritus Professor of Philosophy at the University of Toronto, Mississauga (Canada). Since 2005, his research has been on the philosophy of perception, philosophy of biology and aesthetics. He was elected Fellow of the Royal Society of Canada in 2012.

Aditya Ponkshe is a postdoctoral researcher at the Minimal Intelligence Laboratory (MINT Lab) at the University of Murcia, Spain. He is the author of multiple articles in behaviour, ecology and evolution.

Peter Schulte is Associate Professor of Philosophy at Umeå University (Sweden). He is the author of *Mental Content* (2023) and numerous articles on the philosophy of mind and metaphysics.

Markus Wild is Professor of Philosophy at the University of Basel (Switzerland). He mainly works on animal minds and animal ethics.

The Philosophy of Plant Cognition
Introduction

Peter Schulte, Gabriele Ferretti,
and Markus Wild

1 What Is the Philosophy of Plant Cognition?

Plants are often overlooked in everyday life. While we see animals as individual agents that actively engage with their environment, plants are usually perceived as part of the scenery, like hills, rocks and rivers (or as part of the furniture, if we are indoors). When we look at them more closely, however, we find that plants, too, are active beings. The head of the common sunflower, for example, follows the path of the sun during the day, and returns to its starting position during the night, seemingly in anticipation of the dawn (Vandenbrink et al. 2014). The roots of most plants change their growth direction in response to variations in humidity, salinity, nutrient concentration and the direction of gravity (Muthert et al. 2020). And the tobacco plant reacts to attacks by caterpillars of the tobacco hawk moth by releasing a volatile chemical compound that attracts bugs which feed on the moth's eggs (Kessler and Baldwin 2001).

Such plant activities (or "plant behaviours") are fascinating, and they are the subject of intense empirical study. However, they also raise profound theoretical questions that concern both scientists and philosophers. Are these activities the result of genuinely cognitive processes? Can plants be described as "representing" or "perceiving" their environment? Do they exhibit some kind of intelligence? Might we even have reasons to attribute some form of conscious experience to them? These are questions of great interest, not least because they are intimately connected to central interdisciplinary debates about the *nature* of cognition, representation, intelligence, consciousness and other psychological phenomena, as well as to discussions of general conceptual, methodological and epistemological issues. Hence, these questions constitute the core of what we are calling the "philosophy of plant cognition".

In Section 2 of this introduction, we will provide a brief historical overview of how the interdisciplinary debate about the cognitive capacities of plants has developed in the last two decades and describe the current state of play in areas that are of special relevance for the contributions in this

DOI: 10.4324/9781003393375-1

volume. In Section 3, we will turn to the contributions themselves, summarize them and highlight how they relate to the wider debate.

2 Debating the Cognitive Capacities of Plants: A Brief Overview

In the early 2000s, several plant biologists suggested that plants might have capacities that were usually only attributed to animals. First, Anthony Trewavas argued that plants can "learn" (Trewavas 2001) and that plant behaviour can be described as "intelligent" (Trewavas 2003). A little later, Eric Brenner and several co-authors published an influential article with the explicit aim of establishing the field of "plant neurobiology" (Brenner et al. 2006). In this article, they maintained that plant behaviour is "coordinated across the whole organism by some form of integrated signalling, communication and response system" that is, in crucial respects, analogous to the nervous system in animals (Brenner et al. 2006, 413). Roughly at the same time, the *Society for Plant Neurobiology* was founded (later renamed the *Society of Plant Signaling and Behavior*).[1]

Subsequent years saw a number of interesting developments in this area, of which we can only mention a few. For instance, František Baluška, Stefano Mancuso and their colleagues attempted to revive a version of the "root-brain hypothesis" developed by Charles and Francis Darwin, according to which "the root apex may be considered to be a 'brain-like' organ endowed with a sensitivity which controls its navigation through soil" (Baluška et al. 2009, 1121; see also Baluška et al. 2010). Monica Gagliano and her co-authors presented findings which were supposed to show that *Mimosa pudica* acquires "memories" (Gagliano et al. 2014) and that the pea plant is capable of associative learning (Gagliano et al. 2016). And Richard Karban and John Orrock even proposed a "judgment and decision-making model for plant behavior" (Karban and Orrock 2018).

It is important to note, however, that many of these claims and theoretical proposals are still highly controversial. In reply to Brenner et al. (2006), Amedeo Alpi and 35 other plant biologists questioned the notion of "plant neurobiology", arguing that it "does not add to our understanding of plant physiology, plant cell biology or signaling" (Alpi et al. 2007, 135–6). Both the "root-brain hypothesis" and more wide-ranging claims about plant consciousness were attacked by Lincoln Taiz et al. (2019) and Jon Mallatt et al. (2021). Finally, Gagliano's experiment about associative learning in peas has yet not been successfully replicated and has even failed to replicate in a recent follow-up study (Markel 2020).

While these are, to some extent, controversies within plant biology, questions about the cognitive abilities of plants have also increasingly found their way into the philosophical debate (see, e.g., Calvo 2007; Calvo and Keijzer 2011; Maher 2017; Adams 2018). We will focus here on three major topics of this debate, namely on *cognition, perception* and *consciousness* in plants.

What is the nature of cognition? This is, for obvious reasons, a central question in the philosophy of cognitive science (often discussed under the heading of the "mark of the cognitive"), and it is also of great interest to many cognitive scientists. One way of approaching this question is to begin by identifying a set of paradigmatic cognitive capacities in humans and other animals (the capacity to avoid obstacles, to distinguish between prey and non-prey, to recognize conspecifics, to navigate one's environment, to use tools, to use language, etc.), and then, in a second step, to ask about the nature of the processes that underlie them.[2] When theorists follow this common strategy, however, a deep disagreement emerges. *Representationalists* hold that the relevant processes involve internal representations of environmental and bodily states of affairs, and some form of "manipulation" or "transformation" of these representations.[3] Accordingly, these theorists maintain that explaining the capacities in question partly consists in identifying the representations and operations that are involved in the underlying processes. *Anti-representationalists*, by contrast, reject this view. They either claim that these processes do not involve any internal representations at all (van Gelder 1995; Hutto and Myin 2013)[4] or hold that such representations may well exist but play no significant role in our best explanations of cognitive capacities (Chemero 2009).[5]

This fundamental disagreement has important implications for the question of whether there are genuine cognitive processes in plants. What requirements processes in plants must fulfil to qualify as cognitive, and thus what counts as good evidence for plant cognition, depends heavily on the question of what cognition is. Hence, the question of plant cognition is inextricably bound up with larger questions about the nature of cognitive processes. This point is illustrated, e.g., by the work of Paco Calvo, who argues for the existence of plant cognition from an anti-representationalist viewpoint (Calvo 2016, 1335–7; Segundo-Ortin and Calvo 2019), but also from what is (arguably) a representationalist approach to cognition, namely, the predictive processing approach (Calvo and Friston 2017).[6] Other illustrations are provided by Chauncey Maher (2017), who defends the view that plants have mental capacities on the basis of an anti-representationalist, enactivist framework (inspired by Thompson 2007),[7] and also, in a very different way, by Fred Adams (2018), who draws on a particular version of representationalism to argue against the existence of plant cognition.[8]

Of course, since it is an open question whether plants are cognitive systems, examining the capacities of plants cannot decide the issue between representationalist and anti-representationalist views of cognition, but it brings crucial disagreements into focus and highlights questions that must be answered. For instance, the case of plants pushes representationalists to be more explicit about what conditions a system has to fulfil to count as possessing representational states (a question that is receiving increasing attention, cf. Schulte 2023, 59–63); similarly, it forces anti-representationalists to address the question of what distinguishes systems with cognitive capacities from other systems (since, according to the anti-representationalist, the crucial difference cannot consist in the possession of representational states).

However, the case of plants also raises broader methodological questions. Some theorists hold that the strategy sketched above – to begin with paradigmatic cognitive phenomena in humans and other animals and then to ask whether phenomena of the same kind can be found in plants – exemplifies a problematic form of methodological "zoocentrism" (Figdor 2018; Lyon et al. 2021). Of course, the challenge for these theorists is to develop a convincing alternative approach. Furthermore, this methodological dispute raises the question of whether it is a substantive dispute that reflects a genuine disagreement, or whether it can be resolved by distinguishing between different notions of cognition. In any event, it is clear that plants play a central role in these debates.

When we turn from cognition to perception, similar issues arise. First, there is a fundamental divide between representationalist and anti-representationalist views of perception. Representationalists hold that perceiving objects and their properties involves representing them; they maintain, e.g., that seeing a red sphere on top of a white cube involves visually representing that there is a red sphere on top of a white cube. Anti-representationalists, by contrast, reject this picture and propose to analyse perception without appealing to the notion of representation. Furthermore, representationalists and anti-representationalists also offer very different accounts of the relationship between perception and action. While representationalists typically accept that perception and action are tightly coupled, due to the complex interplay of perceptual and action-oriented representations (for reviews, see Nanay 2013; Ferretti and Zipoli Caiani 2021), many anti-representationalists go further and suggest that the constant interaction between perceiver and environment renders explanatory appeals to internal representations superfluous (O'Regan and Noë 2001; Chemero 2009).

However, there are not just deep disagreements *between* these two camps but also significant differences of opinion *within* them. Different proponents of representationalism offer very different accounts of how perception differs (a) from other forms of representation (most importantly, from beliefs and desires)[9] on the one hand and (b) from non-representational forms of sensory processing on the other, while the proponents of anti-representationalism differ substantially in their views as to what should replace the representationalist paradigm, and how the relationship between perception and action should be construed.

Again, these questions about the nature of perception are closely connected to the question of whether there is perception in plants. Plants are clearly sensitive to numerous environmental stimuli (light intensity, light spectral composition, volatile molecules in the air, salinity, humidity, nutrients in the soil, gravity, etc.) and respond adaptively to them. However, the question is whether these sensory processes are adequately characterized as *perceptual*.

A prominent philosopher who has argued explicitly against plant perception is Tyler Burge (2010, 2014, 2022). According to Burge's theory, a sensory system is perceptual (and thus also representational) if and only if it exhibits perceptual constancies (e.g., size constancy, shape constancy, colour constancy or position constancy).[10] Since the sensory systems of plants do not exhibit such constancies, Burge argues, they do not qualify as perceptual. On his picture, they are paradigm examples of non-perceptual systems of "information registration" (Burge 2014, 392–3).[11] Some theorists have challenged Burge's claim that perceptual constancies are not found in plants (Sims 2019). Others have argued for plant perception on the basis of alternative theories of perception (e.g., Calvo 2016). This illustrates that, just like the question of plant cognition, the issue of whether there is perception in plants is both highly theory-laden and very controversial.

Even more controversial is the last of the three topics that we will discuss here, the topic of consciousness. More precisely, our topic is *phenomenal* consciousness or sentience, the so-called "what-it's-like aspect" of mental states. Philosophers generally agree that there is something it is like for humans to feel pain, to see red or to be scared, and also that there is something it is like to be in these (or similar) states for pigs, dolphins and bats. But is there something it is like to be a daffodil?

When we look at the debate about phenomenal consciousness in general, it is surely no exaggeration to say that it is characterized by deep disagreements about several foundational issues. It is not only that there are different theories on offer that treat very different factors as being crucial for consciousness. It is also that these theories make wildly different predictions about which systems (or system states) are conscious and that their proponents often have fundamentally different conceptions of the explanatory aims and the proper methodology for consciousness research. Many of these disagreements are thrown into sharp relief when we consider the question of plant consciousness.

There are a number of theories of consciousness, e.g., global workspace theories and higher-order theories, which clearly suggest that plants do *not* have conscious experiences. Global workspace theories, for instance, state that a representation becomes phenomenally conscious when it enters a "global workspace", that is, when it is broadcast to a wide variety of consumer systems – e.g., systems for reasoning, decision-making, planning and verbal report (see, e.g., Dehaene and Changeux 2011).[12] Although it is difficult to specify exactly what cognitive structures a system must possess in order to qualify as having a global workspace, it stands to reason that, whatever the details, plants do not have the required structures. With higher-order theories, the case is similar: their core claim is that a representational state is rendered conscious when it becomes the target of a higher-order representation (a "metarepresentation"),[13] and since it is highly unlikely that plants possess metarepresentational capacities, this type of theory also seems to rule out plant consciousness.[14]

On other theories, the implications are less clear. "Integrated information theory" (IIT) is a well-known approach which says that the degree to which a system is phenomenally conscious is given by the information-theoretic quantity Φ (Tononi and Koch 2015). Very roughly, Φ measures "how much information is generated by a system as a whole, compared with its parts considered independently" (Seth and Bayne 2022, 444). Some theorists maintain that IIT supports the claim that plants have conscious experiences (Calvo, Baluška, and Trewavas 2021). However, even if that is true, it should be noted that IIT is highly contentious and has been severely criticized from many sides (Cerullo 2015; Bayne 2018; IIT-Concerned et al. 2023).[15]

Of course, not all arguments concerning the possibility of consciousness in plants are based on full-fledged theories of consciousness. It is also possible to pursue "theory-light" approaches (cf. Birch 2022) with regard to this question. For instance, some theorists have argued for plant consciousness based on considerations of evolutionary continuity and/or by appealing to general functional similarities (see, e.g., Baluška and Reber 2019).[16] However, such arguments are also bound to be controversial (for forceful criticisms, see Brown and Key 2021; Ginsburg and Jablonka 2021; Mallatt et al. 2021). Finally, just as in the case of the concept of cognition, the debate about consciousness or sentience to plants raises the question of whether the conventional use of this concept in cognitive science and philosophy might incorporate superfluous "zoocentric" assumptions.

In this brief survey, we have seen how discussions about cognition, perception and consciousness in plants bring deep disagreements to light. We think that these disagreements, which often involve a blend of empirical, conceptual, methodological and meta-theoretical issues, can only be resolved by combining close attention to the empirical facts with sophisticated philosophical theorizing. Fittingly, this is the general strategy pursued by all contributors to this volume.

3 The Contributions

Part I of this book deals with the topic of cognition and representation in plants. The fundamental question of whether plants exhibit cognitive processes is taken up by Carrie Figdor in her chapter "Why Plant Cognition Is Not (Yet) Out of the Woods". She notes that arguments for plant cognition in the literature usually draw on similarities between plants and animals that are found both on the behavioural level and on the level of the underlying mechanisms. She then identifies two main arguments of this kind. Both arguments start from the premises that (a) plant behaviour seems to be, at a closer look, far more complex than we expected and that (b) this complex behaviour is importantly similar to animal behaviour that we standardly explain by citing cognitive processes. Based on these premises, the Pragmatic Argument proceeds to the conclusion that describing processes in plants as "cognitive" (or "neurobiological") is pragmatically justified because it is scientifically fruitful. The Analogical Argument, on the other hand, proceeds via an inference to the best explanation to the conclusion that plants are, in fact, cognitive systems. Figdor then goes on to subject both arguments to critical scrutiny. The problem with the Pragmatic Argument, in its different versions, is that it is not at all clear whether the metaphorical uses of *neurobiological* terms are, indeed, scientifically fruitful and that establishing the legitimacy of metaphorical uses of *cognitive* terms for plants is not, in general, what plant cognitivists are after. The Analogical Argument, on the other hand, is subject to the "triviality problem" when cognitive terms are understood in a very wide sense, and the "missing phylogeny problem" when they are used in a more restrictive sense. Hence, according to Figdor, we do not currently have a sound basis for the ascription of cognitive capacities to plants.

In "What Deception Reveals about Plant Cognition", Marc Artiga makes a case for the opposing view. He offers a novel defence of "plant cognitivism", the thesis that many plants are cognitive systems. His chapter has two main parts. In the negative part, he argues (against other plant cognitivists) that, taken by themselves, neither the existence of deception in plants nor the existence of plant communication provides strong reasons to accept plant cognitivism. In the positive part, he then suggests that a specific *combination* of communication and deception in plants does provide a forceful argument for this view. According to Artiga, this combination of features (which is in fact found, e.g., in tomato plants) supports the claim that plants are sometimes deceived and can thus misrepresent. Since organisms can only misrepresent if they have representational abilities, and since representational abilities are standardly recognized as cognitive, this leads to the conclusion that plants display cognition. According to Artiga, this result is not only interesting in itself but also suggests particular research strategies and may have important implications for our thinking about the nature of deception.

Peter Schulte, in "Are Plants Representational Systems?", is primarily concerned with the question of whether we can indeed legitimately ascribe representational abilities to plants. First, he argues that the distinction between non-representational and representational systems is thoroughly gradual, and that the position of a system on this spectrum is grounded, roughly speaking, in the degree of information-processing complexity that the system exhibits. Schulte goes on to suggest that the "plant representation thesis", which says that plants are representational systems *simpliciter*, is best interpreted as amounting to the claim that the information-processing complexity in plants is comparable to the complexity we find in insects, fishes and amphibians (which are readily treated as representational systems in cognitive science). On this basis, Schulte examines several types of arguments that have been advanced to support the plant representation thesis. He criticizes "easy arguments" from adaptive responses to variable environmental conditions as fallacious. More sophisticated arguments which draw on physiological similarities between animal brains and electrical signalling systems in plants are also unconvincing, according to Schulte, because they do not succeed in establishing *relevant* similarities. Finally, Schulte argues that arguments based on observations of plant behaviour ultimately fail, too, although there are some intriguing data which suggest that the level of information-processing complexity in plants is *somewhat* higher than previously thought. His conclusion is that we currently do not have good reasons to think that the plant representation thesis is true, and even some reason to doubt that it is.

The next two chapters, which form part II, are concerned with the topic of sensation and perception in plants. In "Not All Sensory Systems Are Information Channels: Outliers from Plant Biology and Beyond", Todd Ganson examines the question of what the nature of sensory systems is, and what we can learn about this question by studying the sensory capacities of plants. According to the traditional view, sensory systems can be defined as "information channels", i.e., as mechanisms that provide the organism with information about its environment. However, Ganson argues that this view is mistaken. First, he draws our attention to an important contrast: while some sensory systems use proximal stimuli as cues for more distal environmental conditions (e.g., shadows as cues for danger), thus enabling the organism to coordinate its behaviour with these distal conditions, other sensory systems do not use proximal stimuli in that way, since their job is to coordinate the organism's behaviour *with these stimuli themselves*. An example of the latter kind of sensory system is the mechanism responsible for positive phototropism in plant shoots: in this case, the direction of light is both the proximal stimulus that triggers the mechanism *and* the condition that the organism needs to coordinate with. In a second step, Ganson then argues that only sensory systems of the first kind can be viewed as information channels and that the traditional view is therefore false as a general account of sensory systems.

Mohan Matthen's chapter, "Plants Sense. But Only Animals Perceive", is also concerned with the sensory systems of plants, but the central question he pursues is whether these systems enable plants to *perceive* their environment. According to Matthen, "perceptions" can be defined as "othering impressions": I perceive an object if I sense it as being distinct from me. Paradigmatic examples of perceptions are visual, auditory and tactile experiences in humans. One crucial feature of these experiences is that they involve a "cognitive map"-like location scheme for external objects, where these objects are represented as standing in certain spatial relations to each other as well as to the perceiver (and thus as clearly distinct from the perceiver herself). Matthen contends that there are no good reasons to postulate such "othering" representations in plants. He acknowledges that the responses of plants to environmental stimuli are quite complex and involve some elementary forms of learning (like habituation), but this alone does not establish that plants perceive. Moreover, he argues that forms of plant behaviour that might be thought to support the claim that plants have perceptual capacities – like obstacle avoidance in plant roots or prey-capture in the Venus flytrap – can, in fact, be explained in simpler ways. By contrast, the behaviour of mammals, birds and possibly fish cannot be accounted for without postulating "cognitive map"-like representations that provide the basis for genuine perception.

The three chapters in part III "Learning, Behaviour and Affordances" take a closer look at the relationship between cognitive and behavioural abilities of plants. The first two chapters do so from a non-representationalist point of view and, more specifically, from within the framework of ecological psychology. By far the most influential version of ecological psychology is James J. Gibson's (1979) theory, which prioritizes the relationship between an animal and its environment, rather than the processing that is going on within an animal's brain. This theory, which can also be described as a theory of direct perception, or perception without mediation by mental representations, is potentially useful for the study of plants. In his chapter "No Brain? No Problem: Toward an Ecological Comparative Psychology", P. Adrian Frazier attempts to lay the groundwork for that approach. Among the key concepts is Gibson's theory of affordances, i.e., of the ways in which an environment can be used by an organism. Trees are approachable and climbable by some vines, for instance. Frazier illustrates this point with examples from the plant literature. Affordances, the author suggests, are a better foundation for comparative psychology than nervous systems, as they constitute a common frame of reference for all creatures, whatever their complexity, whatever their scale. The transition proposed by Frazier is, thus, from plant physiology not to plant neurobiology, but to plant *ecological psychology.*

In "Ecological Plant Learning", Jonny Lee and Aditya Ponkshe also reject the cognitivist assumptions in explaining plant intelligence by suggesting an ecological account of plant behaviour, a central tenet of which is that the correct unit of analysis of a creature's psychology is its environment. They provide an overview of how ecological psychology has hitherto been applied to plants. Focusing on perception, the authors argue that perception is not best explained in terms of isolated sensory stimuli or inferential processes that serve to reconstruct the world to compensate for ambiguous sensory stimuli. Rather, perception must be understood as something that is achieved by the whole organism, embedded in a rich environment with which it is continuously interacting. More precisely, a creature perceives the way it does due to the specifications of its particular surroundings and the creature's own physical properties that are involved in engaging with these surroundings. Applying this to the behaviour of plants, Lee and Ponkshe argue that plants' movements (e.g., the circumnutation of climbing plants) can be explained in the same fashion; plants show ecological anticipatory actions. They go on to discuss whether the ecological-based approach to perception can be extended to encompass learning. What matters for learning, following this approach, is roughly whether there is significant fine-tuning of behaviour via the detection of information in the surroundings of the creature. Lee and Ponkshe argue that, by applying this to plants, ecological psychology can inform our understanding of plant learning by offering a general theoretical framework and by guiding experimental work in a way that involves the rejection of standard cognitivist assumptions about plant intelligence.

In "On Plant Affordances", Gabriele Ferretti is concerned with similar issues. Since plants seem to exhibit a deep interplay between sensory capacities and motor responses, so as to display minimal forms of sensorimotor behaviour, *ça va sans dire* that one interesting line of research would be that of investigating whether plants can perceive affordances. Affordances are usually taken to be crucial in the ecological dimension of cognitive systems, in both human and non-human animals (Gibson 1979). Thus, we may expect that plants, based on their capacity to engage in sensorimotor behaviour, will also be able to detect affordances in the environment. But is this claim so easy to defend? In other words, are we justified in considering plants as capable of detecting affordances just because they exhibit an interplay in sensory encoding and motor behaviour? The task of Ferretti's chapter is exactly to tackle this question. Rather than offering an answer, Ferretti analyses the many complex facets of the notion of affordance and tries to list a series of important questions, each one concerning a single facet, that researchers on plants should address in pursuing the research path that is aiming at defending the possibility of plant affordances. In particular, Ferretti offers insights on the dispositional relation between the biological-cognitive system and the targets it encounters in its ecological niche, a relation from which the affordance seems to emerge, as well as on the information-processing needed for such a system to detect the affordance provided by the target. He analyses all the aspects of these two dimensions of affordances, to prepare the ground for those researchers working on sensorimotor behaviour in plants, who are fascinated by the idea of plant affordances. Ferretti also lists relevant pieces of empirical evidence in plant behaviour, which are those related to climbing, which seem to parallel the case of grasping in humans. Indeed, these may be the candidates that we should put under scrutiny in the light of these considerations on affordances, i.e., with the proviso of taking into account the plethora of constraints, addressed by the author, that affordances posit to those looking for them in the encounter between an animal and its environment.

The final part of this volume concerns the controversial and speculative topic of phenomenal consciousness or sentience in plants. While research on consciousness in animals currently seems to have reached a certain consensus regarding the so-called "markers of conscious experience" in vertebrates, such a position is very controversial and methodologically highly challenging with regard to more disputed sentience candidates such as decapod crustaceans or indeed insects. It seems obvious that this is even more the case with regard to plants. However, it is a consistent pattern in the history of research on non-human organisms that the first hypotheses regarding cognitive abilities are soon followed by speculations about conscious experience. And this is precisely the case with plant consciousness.

In "Making Sense of Plant Sense", Brian Key and Deborah Brown challenge arguments from analogy for plant consciousness that are frequently found in the literature. These arguments aim to establish that plants have conscious experiences from the observation that some of their states perform the same biological function as conscious experiences of animals. While such arguments have been criticized in various ways, Key and Brown focus on a premise that usually goes unchallenged: the assumption that the function of conscious experiences is to cause adaptive behaviour (the "causal assumption"). This assumption is evident, e.g., in arguments from analogy that start from the premise that the function of pain experiences in animals is to trigger avoidance behaviour and/or to teach the animal to avoid potentially harmful actions in the future. Drawing on studies of slugs, fruit flies and mammals, Key and Brown argue that this cannot be the function of pain experiences, since it is a function that is performed by unconscious states in animals. Instead, they suggest that pain experiences, and conscious experiences more generally, have the function of enabling an organism to "make sense" of its (unconsciously caused) behaviour. According to Key and Brown, this not only undercuts arguments from analogy for plant consciousness but also provides us with a positive reason to reject the claim that plants are phenomenally conscious.

In the final chapter of this volume "A Liberal View on Plant Consciousness", Markus Wild argues that the debate on plant consciousness must be conceptually more liberal. As the debate about plant cognition is also an ongoing debate about the concept of cognition, this is equally the case for plant consciousness. More specifically, Wild argues that two features of the standard definition of phenomenal consciousness applied to animals may be dropped. According to the standard definition, for a conscious creature there is something it is like to be that creature, and for a conscious state there is something it is like for that creature to be in that state. However, it is part of this definition that, first, the property of consciousness is a categorical property, since consciousness is an all or nothing affair, and that, secondly, the total conscious state of a creature (creature consciousness) is derivative of individual conscious states (state consciousness) of that creature. Wild rejects these two assumptions, and he argues that consciousness in plants could be either determinate or indeterminate, either dichotomous or degreed, either varied state consciousness or total non-derivative creature consciousness. This liberal view of consciousness is supposed to open up more conceptual space for the debate.

Acknowledgements

The editors of the volume would like to thank the participants and audiences at the conference "Green Intelligence? Debating Plant Cognition" (Basel, October 2021) and the University of Basel, the Swiss National Science Foundation SNSF and the Marie Gretler Stiftung for supporting and funding the conference. Our special thanks go to the anonymous reviewers that commented on the chapters in this volume and to Yannik Steinebrunner (Basel) for preparing the manuscript.

Notes

1 See https://www.plantbehavior.org/about-us/ (last accessed: October 16, 2023).
2 This type of paradigm-oriented approach is defended, e.g., by William Ramsey (2017). It contrasts with approaches that start with an explicit *definition* of "cognition" (e.g., a definition in terms of representation).
3 This is the mainstream view in cognitive science and philosophy of mind. Both "classical" approaches (e.g., Fodor 1975; Pylyshyn 1984) and standard connectionist accounts (e.g., Smolensky 1988; Churchland 2012) are subvariants of it. The earliest clear formulation of the paradigm is probably Craik (1943, ch. 5).
4 To be precise, Hutto and Myin (2013) defend the somewhat weaker claim that internal representations are only involved in processes underlying *linguistic*

abilities. Hence, strictly speaking, their radical enactivism is a "middle position" of the kind described below in note 5.

5 Of course, there is also room for a middle position which asserts that internal representations are required for *some* paradigmatic cognitive capacities, but not for others. See, e.g., Clark and Toribio (1994), who argue that representations are necessary to account for so-called "representation-hungry" capacities.

6 Whether Calvo and Friston actually endorse a form of representationalism is unclear; some passages in their text suggest a representationalist interpretation, while others favor a different reading. Hence, it may either be (a) that Calvo and Friston pursue a (weakly) representationalist approach in parallel to the non-representationalist strategy Calvo adopts elsewhere (in line with the methodological pluralism defended in Calvo 2016, 1334) or (b) that they accept a non-representationalist construal of predictive processing (along the lines of, e.g., Kirchhoff and Robertson 2018).

7 Unlike Calvo and other "plant cognitivists", Maher holds that we do not have strong reasons to attribute representational states to plants (Maher 2017, 94 and 106), so the rejection of representationalism is essential to his case for plant minds.

8 Adams endorses a particularly strong form of representationalism according to which genuine cognition requires the possession of concepts (Adams 2018, 28).

9 Confusingly, this distinction is often characterized as "the perception/cognition divide". This characterization employs a narrow notion of cognition which is different from the wider notion that we are using in this introduction. On the narrow notion, perception is by definition non-cognitive, while on the wider notion, perception is standardly classified as cognitive (or even as paradigmatically cognitive).

10 More precisely, Burge maintains that the presence of perceptual constancies is "*certainly* sufficient for perception and objectivity" and "*conjecture[s]* that they are also necessary" (Burge 2010, 413; our emphasis).

11 According to Burge, this is by no means a verbal issue, since the difference between perceptual and non-perceptual sensory systems is a difference in natural kind.

12 Helpful summaries of this approach are provided by Birch (2022) and Seth and Bayne (2022). For a recent critique of both global workspace and higher-order theories, see Block (2023, 417–44).

13 For an up-to-date survey of higher-order theories, see Carruthers and Gennaro (2023).

14 For another argument against consciousness in plants, based on a different theory of consciousness, see Mallatt et al. (2021, 469–72).

15 Segundo-Ortin and Calvo (2022) also discuss some other theories that might support the hypothesis that plants have consciousness, including quantum-based approaches, predictive processing theories and ecological-enactive approaches.

16 In their paper, Baluška and Reber appeal to the "Cellular Basis of Consciousness" model, but this is not a theory that specifies the features in virtue of which organisms are conscious, but merely the general hypothesis "that the cellular nature of life is inherently linked with consciousness" (Baluška and Reber 2019, sct. 2).

References

Adams, Fred. 2018. "Cognition wars." *Studies in History and Philosophy of Science* 68: 20–30. https://doi.org/10.1016/j.shpsa.2017.11.007.

Alpi, Amadeo, Nikolaus Amrhein, Adam Bertl, Michael R. Blatt, Eduardo Blumwald, Felice Cervone, Jack Dainty, Maria Ida De Michelis, Emanuel Epstein, Arthur W. Galston, et al. 2007. "Plant Neurobiology: No Brain, No Gain?" *Trends in Plant Science* 12 (4): 135–6. https://doi.org/10.1016/j.tplants.2007.03.002.

Baluška, František, Stefano Mancuso, Dieter Volkmann, and Peter Barlow. 2009. "The 'root-brain' Hypothesis of Charles and Francis Darwin." *Plant Signaling & Behavior* 4 (12): 1121–7. https://doi.org/10.4161/psb.4.12.10574.

Baluška, František, Stefano Mancuso, Dieter Volkmann, and Peter W. Barlow. 2010. "Root Apex Transition Zone: A Signalling-Response Nexus in the Root." *Trends in Plant Science* 15 (7): 402–8. https://doi.org/10.1016/j.tplants.2010.04.007.

Baluška, František, and Arthur Reber. 2019. "Sentience and Consciousness in Single Cells: How the First Minds Emerged in Unicellular Species." *BioEssays* 41: 1800229. https://doi.org/10.1002/bies.201800229.

Bayne, Tim. 2018. "On the Axiomatic Foundations of the Integrated Information Theory of Consciousness." *Neuroscience of Consciousness* 4 (1): niy007. https://doi.org/10.1093/nc/niy007.

Birch, Jonathan. 2022. "The Search for Invertebrate Consciousness." *Noûs* 56: 133–53. https://doi.org/10.1111/nous.12351.

Block, Ned. 2023. *The Border Between Seeing and Thinking.* Oxford: Oxford University Press.

Brenner, Eric D., Rainer Stahlberg, Stefano Mancuso, Jorge Vivanco, František Baluška, and Elizabeth Van Volkenburgh. 2006. "Plant Neurobiology: An Integrated View of Plant Signaling." *TRENDS in Plant Science* 11 (8): 413–9. https://doi.org/10.1016/j.tplants.2006.06.009.

Brown, Deborah, and Brian Key. 2021. "Plant Sentience, Semantics, and the Emergentist Dilemma." *Journal of Consciousness Studies* 28 (1–2): 155–83.

Burge, Tyler. 2010. *Origins of Objectivity.* Oxford: Oxford University Press.

Burge, Tyler. 2014. "Perception: Where Mind Begins." *Philosophy* 89 (3): 385–403. http://dx.doi.org/10.1017/S003181911400014X.

Burge, Tyler. 2022. *Perception. First Form of Mind.* Oxford: Oxford University Press.

Calvo, Paco. 2007. "The Quest for Cognition in Plant Neurobiology." *Plant Signaling & Behavior* 2 (4): 208–11. https://doi.org/10.4161/psb.2.4.4470.

Calvo, Paco. 2016. "The philosophy of plant neurobiology: a manifesto." *Synthese* 193: 1323–43. https://doi.org/10.1007/s11229-016-1040-1.

Calvo, Paco, František Baluška, and Anthony Trewavas. 2021. "Integrated Information as a Possible Basis for Plant Consciousness." *Biochemical and Biophysical Research Communications* 564: 158–65. https://doi.org/10.1016/j.bbrc.2020.10.022.

Calvo, Paco, and Karl Friston. 2017. "Predicting Green: Really Radical (Plant) Predictive Processing." *Journal of the Royal Society Interface* 14: 20170096. https://doi.org/10.1098/rsif.2017.0096.

Calvo, Paco, and Fred Keijzer. 2011. "Plants: Adaptive Behavior, Root-Brains, and Minimal Cognition." *Adaptive Behavior* 19 (3): 155–71. https://doi.org/10.1177/1059712311409446.

Carruthers, Peter, and Rocco Gennaro. 2023. "Higher-Order Theories of Consciousness." In *The Stanford Encyclopedia of Philosophy* (Fall 2023 Edition), edited by Edward N. Zalta and Uri Nodelman, URL = https://plato.stanford.edu/archives/fall2023/entries/consciousness-higher/.

Cerullo, Michael A. 2015. "The Problem with Phi: A Critique of Integrated Information Theory." *PLoS Computational Biology* 11 (9): e1004286. https://doi.org/10.1371/journal.pcbi.1004286.

Chemero, Anthony. 2009. *Radical Embodied Cognitive Science*. Cambridge, MA: MIT Press.

Churchland, Paul. 2012. *Plato's Camera*. Cambridge, MA: MIT Press.

Clark, Andy, and Josefa Toribio. 1994. "Doing without Representing?" *Synthese* 101: 401–31. https://doi.org/10.1007/BF01063896.

Craik, Kenneth. 1943. *The Nature of Explanation*. Cambridge: Cambridge University Press.

Dehaene, Stanislas, and Jean-Pierre Changeux. 2011. "Experimental and Theoretical Approaches to Conscious Processing." *Neuron* 70: 200–27. https://doi.org/10.1016/j.neuron.2011.03.018.

Ferretti, Gabriele, and Silvano Zipoli Caiani. 2021. "Between Vision and Action: Introduction to the Special Issue." *Synthese* 198 (Suppl 17): S3899–S3911. https://doi.org/10.1007/s11229-019-02518-w.

Figdor, Carrie. 2018. *Pieces of Mind*. Oxford: Oxford University Press.

Fodor, Jerry. 1975. *The Language of Thought*. Cambridge, MA: MIT Press.

Gagliano, Monica, Michael Renton, Martial Depczynski, and Stefano Mancuso. 2014. "Experience teaches plants to learn faster and forget slower in environments where it matters." *Oecologia* 175: 63–72. https://doi.org/10.1007/s00442-013-2873-7.

Gagliano, Monica, Vladyslav V. Vyazovskiy, Alexander A. Bobély, Mavra Grimonprez, and Martial Depczynski. 2016. "Learning by Association in Plants." *Scientific Reports* 6: 38427. https://doi.org/10.1038/srep38427.

Gibson, James J. (1979). The Ecological Approach to Visual Perception. Boston: Houghton Mifflin.

Ginsburg, Simona, and Eva Jablonka. 2021. "Sentience in Plants: A Green Red Herring?" *Journal of Consciousness Studies* 28 (1–2): 17–33.

Hutto, Daniel, and Erik Myin. 2013. *Radicalizing Enactivism*. Cambridge, MA: MIT Press.

IIT-Concerned, Stephen M. Fleming, Chris D. Frith, Melvyn A. Goodale, Hakwan Lau, Joseph LeDoux, Alan L.F. Lee, Matthias Michel, Adrian M. Owen, Megan A.K, et al. 2023. "The Integrated Information Theory of Consciousness as Pseudoscience." *PsyArXiv Preprints*: https://doi.org/10.31234/osf.io/zsr78.

Karban, Richard, and John L. Orrock. 2018. "A Judgment and Decision-Making Model for Plant Behavior." *Ecology* 99 (9): 1909–19. https://doi.org/10.1002/ecy.2418.

Kessler, André, and Ian T. Baldwin. 2001. "Defensive Function of Herbivore-Induced Plant Volatile Emissions in Nature." *Science* 291: 2141–4. https://doi.org/10.1126/science.291.5511.2141.

Kirchhoff, Michael, and Ian Robertson. 2018. "Enactivism and Predictive Processing: A Non-Representational View." *Philosophical Explorations* 21 (2): 264–81. https://doi.org/10.1080/13869795.2018.1477983.

Lyon, Pamela, Fred Keijzer, Detlev Arendt, and Michael Levin. 2021. "Reframing cognition: getting down to biological basics." *Philosophical Transactions of the Royal Society B* 376: 20190750. https://doi.org/10.1098/rstb.2019.0750.

Maher, Chauncey. 2017. *Plant Minds: A Philosophical Defense*. London: Routledge.

Mallatt, Jon, Michael R. Blatt, Andreas Draghun, David G. Robinson, and Lincoln Taiz. 2021. "Debunking a Myth: Plant Consciousness." *Protoplasma* 258: 459–76. https://doi.org/10.1007/s00709-020-01579-w.

Markel, Kasey. 2020. "Lack of evidence for associative learning in pea plants." *eLife* 9: e57614. https://doi.org/10.7554/eLife.57614.

Muthert, Lucius Wilhelminus Franciscus, Luigi Gennaro Izzo, Martijn van Zanten, and Giovanna Arrone. 2020. "Root Tropisms: Investigations on Earth and in

Space to Unravel Plant Growth Direction." *Frontiers in Plant Science* 10: 1–22. https://doi.org/10.3389/fpls.2019.01807.

Nanay, Bence. 2013. *Between Perception and Action*. Oxford: Oxford University Press.

O'Regan, J. Kevin, and Alva Noë. 2001. "A Sensorimotor Account of Vision and Visual Consciousness." *Behavioral and Brain Sciences* 24: 939–1031. https://doi.org/10.1017/s0140525x01000115.

Pylyshyn, Zenon. 1984. *Computation and Cognition: Toward a Foundation for Cognitive Science*. Cambridge, MA: MIT Press.

Ramsey, William. 2017. "Must Cognition Be Representational?" *Synthese* 194: 4197–214. https://doi.org/10.1007/s11229-014-0644-6.

Schulte, Peter. 2023. *Mental Content*. Cambridge: Cambrigde University Press.

Segundo-Ortin, Miguel, and Paco Calvo. 2019. "Are plants cognitive? A reply to Adams." *Studies in History and Philosopy of Science* 73: 64–71. https://doi.org/10.1016/j.shpsa.2018.12.001.

Segundo-Ortin, Miguel, and Paco Calvo. 2022. "Consciousness and Cognition in Plants." *WIREs Cognitive Science* 13: e1578. https://doi.org/10.1002/wcs.1578.

Seth, Anil K., and Tim Bayne. 2022. "Theories of Consciousness." *Nature Reviews Neuroscience* 23: 439–52. https://doi.org/10.1038/s41583-022-00587-4.

Sims, Matthew. 2019. "Minimal Perception: Responding to the Challenges of Perceptual Constancy and Veridicality with Plants." *Philosophical Psychology* 32 (7). https://doi.org/10.1080/09515089.2019.1646898.

Smolensky, Paul. 1988. "On the Proper Treatment of Connectionism." *Behavioral and Brain Sciences* 11: 1–74. https://doi.org/10.1007/978-94-009-1882-5_6.

Taiz, Lincoln, Daniel Alkon, Andreas Draghun, Angus Murphy, Michael Blatt, Chris Hawes, Gerhard Thiel, and David G. Robinson. 2019. "Plants Neither Possess nor Require Consciousness." *Trends in Plant Science* 24 (8): 677–87. https://doi.org/10.1016/j.tplants.2019.05.008.

Thompson, Evan. 2007. *Mind in Life*. Cambridge, MA: Harvard University Press.

Tononi, Giulio, and Christof Koch. 2015. "Consciousness: Here, There and Everywhere?" *Philosophical Transactions of the Royal Society B* 370: 20140167. https://doi.org/10.1098/rstb.2014.0167.

Trewavas, Anthony. 2001. "How Plants Learn." *Proceedings of the National Academy of the Sciences* 96: 4216–8. https://doi.org/10.1073/pnas.96.8.4216.

Trewavas, Anthony. 2003. "Aspects of Plant Intelligence." *Annals of Botany* 92: 1–20. https://doi.org/10.1093/aob/mcg101.

van Gelder, Tim. 1995. "What Might Cognition Be, If Not Computation?" *Journal of Philosophy* 92 (7): 345–81. https://doi.org/10.2307/2941061.

Vandenbrink, Joshua P., Evan A. Brown, Stacey L. Harmer, and Benjamin K. Blackman. 2014. "Turning heads: The biology of solar tracking in sunflower." *Plant Science* 224: 20–6. https://doi.org/10.1016/j.plantsci.

Part 1
Cognition and Representation

1 Why Plant Cognition Is Not (Yet) Out of the Woods

Carrie Figdor

1.1 Introduction

There is no doubt that plants and animals share genetic, molecular, developmental, morphological and behavioral traits and that plants exhibit flexible, adaptive behavior. The question is whether these similarities support the conclusion that plant signaling and behavior is best explained by ascribing cognitive processes to them. In this chapter I present the two main arguments advanced for plant cognition and find both of them very weak. Neither challenge rests on an exclusionary definition of "cognition" nor on an anthropocentric or zoocentric perspective. They rely on the contemporary biological research context and general features of biological trait individuation. I'll present the main arguments in Section 1.2 and critically assess them in Sections 1.3 and 1.4.

1.2 The Main Arguments for Plant Cognition and Intelligence

Plant-cognition advocates (advocates, for short) have provided extensive discussions of new findings in plant science to support their case for considering adaptive, flexible plant behavior as cognitive and intelligent (e.g. Calvo 2016; Segundo-Ortin and Calvo 2019, 2022 ; Calvo et al. 2020); Calvo and Keijzer 2009, 2011; Trewavas 2003; Baluška and Mancuso 2009; Baluška, Mancuso, and Volkmann (eds.) 2006). An overlapping debate concerns whether plant physiology should be classified within neurobiology (e.g. Baluška and Mancuso 2007; Trewavas 2007). The possibility of plant consciousness has recently been added to plant advocacy (advocacy, for short) as similar evidence is used to support both conclusions (e.g. Calvo 2017; Segundo-Ortin and Calvo 2023; Mediano, Trewavas, and Calvo 2021). Critics target at least one of the claims in this advocacy mix (Adams 2018; Firn 2004; Alpi et al. 2007; Chamovitz 2018; Birch 2023; ten Cate 2023). I will focus here on the arguments for plant cognition; it will fall out of my criticisms that arguments for plant consciousness must be assessed independently.

DOI: 10.4324/9781003393375-3

The main line of support for plant cognition depends on similarities of behavior and mechanisms in animals and plants. "Mechanisms" includes the genetic, molecular, anatomical or other physiological mechanisms that are homologous or analogous between animals and plants. Both behavior and mechanisms are catalogued exhaustively in the papers referenced above and are omitted here for space reasons. These empirical features are used to argue for plant cognition as follows:

(P1) The adaptive behavior of at least some extant angiosperms and gymnosperms are more flexible and complex than many have thought.

(P2) This behavior, along with many key supporting internal signaling mechanisms, are robustly similar to those of animals.

(P3) We explain this adaptive, flexible behavior in the case of animals by citing the intelligent, cognitive processes that cause them.

(C1) The pragmatic conclusion: Calling (C1a) the plant behavior "intelligent" and "cognitive", and (C1b) the supporting mechanisms "neurobiological" is justified because the use of these terms, however metaphorical, is scientifically fruitful. While both labels have long been restricted to animals, using them for plants promotes more (or maybe better) research by promoting a more "open attitude" among plant scientists (e.g. Calvo and Keijzer 2011, 156).

(C2) The analogical/IBE conclusion: The best explanation of the complex, flexible behavior of plants is that it too is caused by cognitive, intelligent processes (Segundo-Ortin and Calvo 2019).[1]

For clarity, I will call the argument ending in the pragmatic conclusion (C1) the pragmatic argument and the argument ending in the analogical/IBE conclusion (C2) the analogical/IBE argument, even though both rely on the same premises and analogical inference. These are the two main arguments. Many advocates also promote plant cognition through redefinitions that assimilate "intelligent", "flexible" (or "non-hardwired") and "cognitive" over the course of an article. This type of argument by redefinition goes roughly as follows:

(RP1) If behavior is flexible and adaptive, then it is not hardwired.

(RP2) Definition: not hardwired = intelligent.

(RP3) Plant behavior is flexible and adaptive.

(RC1) Therefore, plant behavior is intelligent.

(RP4) Intelligence includes capacities that cognitive science considers cognitive processes, such as selecting actions to achieve goals and exhibiting anticipatory behavior.

(RC2) Therefore plants have cognitive processes (are cognitive).[2]

Briefly, "flexible" = "adaptive" = "intelligent" = "cognitive" in many advocates' papers, such that acknowledging flexibility in plant behavior is tantamount to acknowledging plant cognition. Although I will set this redefinitional argument aside, definitions of cognition in cognitive science play a key and legitimate role in the analogical/IBE argument and are discussed below.

The truth of (P1) is not in dispute. Some of the key empirical results aren't new – in particular, action-potential-based cell-to-cell signaling (Bose 1926) and habituation (Holmes and Gruenberg 1965) – but many more plant genetic, anatomical, physiological and developmental details are now known, and more experiments on plant learning at the level of whole plants or their parts are being conducted, particularly using designs adapted from comparative psychology (Abramson and Chicas-Mosier 2016). The jury is very much still out on associative learning (e.g. Gagliano 2016; Markel 2020). But there is no doubt that at least some plants exhibit non-associative learning; therefore, they learn in some specific ways that are ascribed to animals, and at least some plants have cell-to-cell action-potential-based signaling systems; therefore, they also have at least one key mechanism known to subserve cognitive processing in animals.

(P2) and (P3) and both conclusions will be discussed below. However, the basic issue is what inferences we are justified in drawing from the new or reinforced facts about plant behavior and mechanisms. Should we conclude that plants make decisions in the same sense animals do, anticipate future contingencies in the same sense animals do, and so on?

1.3 The Pragmatic Argument

To address the pragmatic argument, it suffices to consider only whether the use of cognitive language for plants is justified by its scientific utility even if it is merely metaphorical. Here we need to distinguish (C1b) from (C1a).

The neurobiological language (C1b) is acknowledged to be metaphorical. Trewavas (2007) argues that these metaphors are scientifically valuable because they reveal experimental questions that might not otherwise be immediately obvious. Segundo-Ortin and Calvo (2022, 8) concur, adding that, "the price of using metaphors and analogies is eternal vigilance". Obviously, the vigilance isn't eternal. Even useful metaphors are eventually cashed out with knowledge of actual mechanisms that provides non-metaphorical ways of referring to them. But the pragmatic argument credits the metaphors with raising new experimental questions in plant science, or at least with raising them sooner rather than later. This claim is surprisingly weak. There is ample reason to think these metaphors have played no useful scientific role whatsoever.

For one thing, some metaphors or analogies have rolled around for a long while without any obvious positive effect on science. They may be used in implicit arguments from authority (e.g. the many references to Darwin's (1880) "root brain") or may be hindrances to plant research (Tompkins and Bird 1974).[3] Darwin (1859, 523, 525) himself noted that analogy can be "deceitful" and considered progress in the natural sciences to include the elimination of metaphors (while frankly acknowledging his own metaphorical uses).

More importantly, genetics, genomics and development (Evo-Devo) have transformed biological research across the board, including in plant science (e.g. Olson 2019). Evo-Devo in particular has made clear how misguided and misleading the fixed/innate vs. adaptive/learned dichotomy is for all organisms, in particular multicellular ones. Biology as a whole has moved away from framing nonhuman organism behavior as inflexible and innate. So has comparative psychology and its cognate sciences since the loosening of behaviorist and early ethologists' strictures. Even "instincts" require appropriate input and interaction with environment (Burghardt and Bowers 2017). In short, the context of contemporary biology and comparative psychology provides sufficient reason to reject as false an old dichotomy in which plants are inflexible and (at least some) animals are not, independently of the use of neurobiological metaphors.

There is also sufficient motivation from phylogeny to ask and investigate comparative questions across the plant and animal kingdoms, simply because *any* claim about a species or clade entails questions and suggests hypotheses about any other, whether phylogenetically close or distant. What is new is our ability to investigate these questions at multiple mechanism levels at high resolution. For example, we know that MADS box and homeobox gene families are homologous in plants and animals – the same genes in each, due to inheritance from their last common ancestor – but that the roles of these genes (as master transcription regulators) in plant and animal development are not homologous (Meyerowitz 2002). We are continuing to precisely characterize the multiple roles of the hormone auxin in angiosperm development and growth (Finet and Jaillais 2012). In this case, advocates mischaracterize auxin as a "neurotransmitter" (Baluška et al. 2006, 19; Calvo and Keijzer 2009, 257), let alone one transported by "metaphorical neurons" (Trewavas 2007, 232). That language misleadingly implies that auxin has a specific (and brain-ish) function, when multifunctionality is common throughout biology (as is degeneracy).

The non-metaphorical similarities between plant and animal mechanisms and behavior in (P1) have also led to proposals for conceptual revision that aims to assimilate all these new empirical details. For example, Miguel-Tomé and Llinás (2021) propose an expanded definition of a nervous system so that plant cell-to-cell signaling systems count as nervous systems. But they do not propose to redefine the neuron character as well: differences between plant and animal signaling cells remain significant even while these and other comparisons bring into relief new research questions (Moroz and Romanova 2021, 2022; Burkhardt et al. 2023). Similarly, Cvrckova et al. (2016) propose a definition of behavior intended to encompass both plant growth and animal motion: "observable consequences of the choices a living entity makes in response to internal or external stimuli". But they also caution (2016, 3) that "choice" in their definition is just a matter of adopting one of at least two alternative trajectories in the available state-space: "by no means does this word imply involvement of a mind or consciousness".[4] Their definition also excludes artificial systems; the significance of this will be seen in Section 4.

Both cases of proposed conceptual revision involve defining a superordinate category that includes plants and animals without abandoning important distinctions between them that are key for promoting precise testable hypotheses for future research. Accumulated facts about plants and comparative work with other species suffice to motivate these proposals. Such revisions are a regular feature of science after more facts have accumulated that need to be integrated into phylogeny. Now we divide the tree of life into three domains descended from the Last Universal Common Ancestor (LUCA), given the genetic and molecular distinctness of Archaea from both Bacteria and Eukarya (Woese et al. 1978). Now it is standard to distinguish hominins from hominids in the light of discoveries of many extinct *Homo* species. And so on: more empirical facts, more or updated categories to organize them in phylogeny.

In summary, general advances in biology have clearly promoted new empirical research and theoretical progress in plant science. Whatever the rhetorical effects of neurobiological metaphors, it is highly dubious to credit them with any scientific utility. They seem at best epiphenomenal, at worst misleading.

How about (C1a): is the *cognitive* terminology also metaphorical? Sometimes advocates are not entirely clear (Calvo 2017, 220, fn. 10); their fellow travelers and occasional collaborators in bacterial, basal or minimal cognition are more often equivocal or ambiguous (Lyon 2015, 2; Lyon et al. 2021; Levin et al. 2021). In any case, in the absence of unequivocal claims that it is metaphorical – for example, explicit claims that plants behave *as if* they made decisions – the charitable interpretation is that advocates do not intend to use the cognitive language metaphorically. Thus, I disagree with Birch (2023) that plant "decisions" and so on are metaphorical, even though we agree the cognitive language is problematic. A non-metaphorical interpretation ensures that the analogical/IBE argument is still worth making. No one uses an analogical/IBE argument to conclude that other people act *as if* they make decisions. This interpretation also explains why advocates go great lengths to provide definitions of intelligence and specific cognitive abilities that are inclusive of plants and to show that these definitions are not *ad hoc*. They aren't. But they don't help the cause. In a word, while advocates legitimately use standard cognitive-scientific concepts to provide cognitive explanations of flexible plant behavior, our best explanations of that behavior will rely on a distinct cognitive-conceptual framework based in evolutionary biology (introduced in Section 4). For all we know now, plants may not be cognitive in that framework.

1.4 The Analogical Argument

Segundo-Ortin and Calvo (2022, 3) state that "positing cognitive processes in plants is an inference to the best explanation of what can possibly underlie their behavioral repertoire". The similarities between plants and animals included in (P1) are "robust enough" to justify plant cognition.[5] Similarly, Calvo and Keijzer (2011, 156) claim that "the intricacy of the underlying organizations warrants the use of a cognitive terminology here". In an interesting new twist, Lee, Segundo-Ortin, and Calvo (2023) argue for plant decision-making by analogy to *bacteria* cognition; this builds on Lyon (2015, 3), who suggests an analogical argument between bacteria and animals. Thus is a house of cards constructed.

Of course, the specific similarities in (P1) don't need a qualitative assessment of robustness. Auxin hormones are the same in plants and animals even if their functional roles differ in each. Habituation is the same form of learning in both plants and animals. In other terms, "auxin" and "habituation" are not equivocal when used in reference to plants and animals. The question is whether we are justified in concluding from these homologies or homoplasies that plants make decisions in the same sense animals do and so on. It is question-begging to support this conclusion by citing other advocates who describe plant behavior using cognitive terms.

I will raise two major problems for the analogical argument. I introduce them briefly now. The first is that advocates succeed in showing that plants and animals have the same cognitive abilities only when the abilities ascribed in the argument are the cybernetic-computational abilities shared across biological and artificial systems alike. So it is not surprising that we can treat plants in particular as cybernetic-computational systems. I call this the triviality problem. The second is that we are not entitled to infer that the best explanation of the similarities is that plants have cognitive abilities in any more restricted sense, because we don't yet have the cognitive-conceptual tools we need to provide the explanations of biological behavior we want. It is completely an open question whether the best explanation of plant behavior will require using those conceptual tools when we have them. I call this the missing phylogeny problem.

The triviality problem. This problem stems from mainstream cognitive science, which via cybernetics ushered in the use of such concepts as goals, decisions or choices, anticipation and so on for systems that have feedback control of behavior, whether biological or artificial (Wiener 1948/2019; Milkowski 2018-; Figdor 2018b, 2023; Bassel 2018). This cybernetic-computational framework applies quite widely, just as Wiener intended: *qua* cybernetic-computational (CC) systems, animals, bacteria, robots, self-supervised neural networks, cells, organs and so on are conceptually all of a piece.[6] The philosophical cover for this generality is the principle of medium-independence, adopted from traditional functionalism in philosophy of mind (explicitly endorsed in Levin et al. 2021, 2–3). So does (P1) justify the claim that plants are like animals in that plants too are CC systems and can be ascribed the abilities CC systems have?

Well, part of (P1) doesn't do any work in supporting this claim. From the CC perspective, advocates shouldn't care whether plant mechanisms are similar to those of animals, given the medium-independence of the framework. Even if there were no similarities in plant and animal mechanisms, plants could *still* make decisions in the CC sense; even if plant mechanisms were more similar to those of bacteria than animals, the same conclusion would follow. In any event, the qualitatively "robust" similarities in mechanisms in (P1) are irrelevant once cognition and intelligence are understood in traditional cognitive-scientific terms.

In contrast, the behavioral similarities in (P1) are essential. They show that plant behavior and signaling are adaptive and flexible, in contrast to being done in a "fixed hard-wired manner" (Segundo-Ortin and Calvo 2022, 3). Their flexibility is shown by the facts that (at a minimum) they learn by sensitization and habituation and grow in ways that are responsive (usually adaptive) to their environments. So we are entitled to conclude that plants have at least the flexibility of an autonomous cybernetic system. We can subsume the processes underlying their flexible, adaptive behavior within the CC framework and use that framework's vocabulary of goals, decisions, learning, memory and so on to refer to what plants and animals share, qua CC systems, without equivocation. Advocates are completely justified in talking about plant cognition when we adopt this view of organisms as information-processing systems embedded in and flexibly responding to their environments in goal-directed ways – that is, as CC systems.

However, this extension of the CC framework to plants is not at all surprising in the context of our general rejection, noted above, of the idea that nonhuman biological organisms act in fixed, hardwired ways. We should not forget that until very recently many animals were no better off than plants in this respect (e.g. Keijzer 2013). Now, besides plants, bacteria and archaea have their champions (Lyon 2015; Baluška and Levin

2016; Shapiro 2021) and Team Fungi has recently joined the stramash (e.g. Adamatzky et al. 2023; Money 2021). Recognition of this feedback control of behavior (a.k.a. "adaptive flexibility") means we can use computational or mathematical models and methods to model plant behavior. That is interesting and important. But using mathematical or computational models for biological organisms is the easy part. The hard part is understanding what these models tell us about real organisms and what the relationship is between the models and the organisms they model. That is a bigger issue than I can address here (see, e.g. Chirimuuta 2021). For current purposes, however, it is sufficient to recognize that the abilities ascribed in these toy models are toy abilities with no essential relationship to biology. They are no more and no less than those we also ascribe to robots, self-driving cars and other advanced autonomous AIs. The same goes for cognitive explanations of at least some *animal* behavior.[7] The CC framework is available for use for any autonomous feedback-control system.

This is easy to forget when the CC vocabulary is grafted onto the pre-existing Darwinian vocabulary of adaptive responsiveness in the service of survival and reproductive goals that governs all biological organisms. This assimilation makes the CC and Darwinian vocabularies either synonymous within the biological domain (Cvrčková, Lipavská, and Žárský 2009, 394) or homonymous; at the very least, there is very little light between them. Thus, although (e.g.) Calvo and Keijzer (2011, 169) define cognition from what they call a "biocentric" perspective as "an ability to manipulate the environment in ways that systematically benefit a living organism", there is nothing essentially biocentric about the CC framework that entitles them to use cognitive terminolgy. To the contrary, it is essentially non-biological. This is why an argument from analogy to plant consciousness must be assessed independently of the argument to plant cognition. The CC framework omits consciousness and because of this has long been accused of not capturing biological minds. So the mechanisms in (P1) may do work in that argument that they can't do in this one, given this omission.[8]

It is illuminating to contrast this unproblematic extension of CC concepts to plants (*inter alia*) with the inappropriateness of dragging in concepts from psychology without care. For example, Trewavas (2003, 2005, 2016; see also Calvo et al. 2020) takes definitions of "intelligence" from psychology (e.g. a capacity for problem-solving, an ability to learn from experience) at face value and argues that they also apply to plants. But the traditional conceptual anthropocentrism of psychology guarantees fallacies of equivocation if these terms are extended to nonhumans without semantic adjustment of some kind. Bennett and Hacker (2003) argue on just these grounds that uses of psychological terms for nonhumans are *nonsense*. I disagree (Figdor 2018a), but the point remains that many ordinary psychological concepts are not straightforwardly extendable to nonhuman organisms the way CC concepts clearly are.

The missing phylogeny problem. This problem is fatal. The fact that biological organisms can be treated as CC systems does not entail that their behavior is best explained in terms of the CC framework and the concepts defined within it. If we are interested in explaining plant behavior and not just extending cybernetics to plants, we need to identify and individuate the cognitive processes whereby, for any plant species that has them, *being a plant of that species* matters for having them. The missing phylogeny problem is that we are not entitled to infer (C2) if (C2) is a claim about the behaviors and cognitive abilities of plants *qua* plants. It is entirely an open question whether the best explanation of plant behavior we eventually give will involve ascribing cognitive characters to them. For the time being, plant cognition is pure speculation if it is not cybernetics.

The best cognitive explanation of the flexible behavior of any biological organism will require at least the start of a conceptual framework of cognitive characters. To a first approximation, (phylogenetic) characters are inherited traits individuated across species; they capture that aspect of a species' phenotypes that are due to the species' evolutionary history. They are individuated at every level of biological organization, from genes to behavior. The tree of life – phylogeny – is created based on shared characters that form a nested hierarchy of clades (monophyletic groups of species, which include an ancestral species and all and only its descendant species). They are critical for making inferences about other inherited features that might also be shared across species, including species that have not been directly examined. Characters are also mapped to phylogenies as well as creating them (e.g. acoustic communication across major animal clades) (Chen and Wiens 2020). Behavioral characters are of this sort, and cognitive characters will almost certainly follow suit.

Character individuation is a complicated matter in which similarities and differences at multiple levels are taken into account.[9] This is because distinct species' phenotypes that determine shared characters can be dissimilar in form or function (e.g. human and narwhal canine teeth), and phenotypes that determine distinct characters can be similar in form or function (e.g. shark and dolphin fins). Because characters (unlike phenotypes) are individuated across species, individuating them depends essentially on interspecies comparisons – the more species, the better. Since every species has its place in phylogeny, the phylogenetic positions of the compared species matter. If a new species has or is ascribed a cognitive ability, its behavioral phenotypes must count as evidence of that ability and may provide reason to revise the initial definition of the ability. This is an iterative process that depends partly on what else we know about the newly ascribed species, including its position in phylogeny. All this means that similarities in (P1) *and* differences *not* in (P1) *and* comparisons with other species throughout phylogeny will be examined when individuating cognitive characters.

It also follows that adding just one new species in a new kingdom or other major clade to the extension of a character is a Very Big Deal: it has implications for the individuation of that character, its ascription to other new species, our inferences to other characters species that have it might also have, and our best explanation of its origin and extent in phylogeny. The latter is the key issue here, since the plant cognition debate fundamentally concerns our best hypotheses regarding when each cognitive ability initially appeared in phylogeny and how it differentiated and spread throughout phylogeny. In other words, once biology matters, the debate is not just about plants.

For example, we know that plants and animals both learn by habituation (a comparative biological/psychological concept): "habituation" picks out the same behavioral character across these lineages despite differences in behavioral and internal mechanism phenotypes. We might ask if plants and animals both habituate because their last common ancestor could learn by habituation (homology) or because habituation developed in both lineages independently due to common environmental pressures (homoplasy). We know neurons are not necessary for habituation, although we don't know yet what neurons specifically are needed for. Differences between neuron-based and non-neuron-based signaling systems may lead us to develop different habituation subtypes exhibited by plants and animals. Maybe learning by association is limited to neuronal signaling, but animals and plants share some components. This might motivate either revising its definition or defining new component characters. In all cases the guiding aim is to organize the behavioral and cognitive complexity we are discovering throughout phylogeny into a conceptual framework of characters that capture the cross-species similarities most important for explaining organism behavior (Delaux et al. 2019). Whether some advocates judge some similarities "robust enough" is entirely beside the point. (P1) very likely contains a mix of homologies, homoplasies and superficial similarities, some of which may be important for individuating cognitive characters and some of which won't matter.

Notably, the medium-independence foundational to cognitive science and the CC conceptual framework entails that CC concepts do not pick out characters. Phylogeny makes no difference to the individuation of CC concepts. That is why adding a whole new biological kingdom to the extension of a CC concept can be done without blinking an eye. But what the CC framework explicitly ignores is essential to character individuation. Even the proposed redefinitions of nervous systems and behavior mentioned above exclude AI systems. Taking into account the amassed plant data motivated these researchers to rethink these concepts, but they are still thinking of them in biological, hence phylogenetic, terms.[10]

At the same time, merely extending a CC concept to a biological organism does not *turn it into* a character. First, from a biological perspective we can't plausibly generalize to untested species in a new kingdom or major clade if a small sample of species in that kingdom or clade has a character. This is especially the case if we think the newly sampled species have it through convergent evolution rather than by inheriting it from their last common ancestor with species in a distinct kingdom or major clade that share the character. From a phylogenetic perspective, the number of plant species that have actually been tested is miniscule. Inferences to all plant species are entirely speculative. None of this is problematic if the cognitive concepts extended to plants are not characters. The fact that we've only tested a few plant species is irrelevant for extending CC concepts to all of them (assuming all are adaptively flexible).

Second, CC concepts used in the biological domain have none of the explanatory power of characters. They provide no information about the evolutionary relationships of organisms or their cognitive traits. This is not surprising, as it doesn't matter how species are related for them to have the same CC cognitive ability. Although evolutionary biologists have generally preferred homologous characters, even homoplasous characters – characters that are shared across species for reasons not due to common ancestry – have explanatory import, such as for identifying common evolutionary pressures across major clades and suggesting possible homologies at other levels of biological organization (e.g. shared mechanisms of some sort). But homoplasy and homology occur at every level of biological organization. In particular, behavioral characters may be homologous or homoplasous – it depends on the behavior (e.g. de Queiroz and Wimberger 1993; Rendall and DiFiore 2007) – and we can expect the same for cognitive characters. In contrast, there is no principled way within the CC framework to give different explanations of why distinct species have the same CC cognitive abilities; we can't even ask why distinct species have different (CC) cognitive abilities. Breathtaking conceptual generality has its uses, but it is hamstringing when we are just beginning to investigate the cognitive abilities of a wide variety of nonhumans and hypothesize when and how various cognitive abilities emerged and evolved. This research depends essentially on comparisons across multiple species and taxa. The CC framework doesn't enable us to draw the distinctions and offer the explanations that are just now coming within reach.

Ironically, cognitive-conceptual anthropocentrism blocked traditional cognitive concepts from being used to pick out biological groups based on superficial similarities (the way "fish" once included whales). In an anthropocentric conceptual scheme, cognitive concepts are defined by human cognitive phenotypes; these concepts are then used as prototypes similarity to which determines whether a nonhuman entity has a cognitive ability. This historically dominant view is phylogenetically bankrupt. But in their eagerness to bury anthropocentrism, advocates jump from anthropocentric cognitive concepts to cognitive concepts that group organisms regardless of evolutionary origin or relationships. They reclassify plants from being not cognitive in an anthropocentric conceptual framework to being cognitive in a cybernetic-computational conceptual framework. This perpetuates a false conceptual dichotomy of either cybernetic or anthropocentric cognitive concepts and abilities (e.g. Lyon (2015, 3, Table 1) at a time of burgeoning research into the phylogeny of behavior and cognition. Advocates show awareness of this lacuna between anthropocentrism and cybernetics (e.g. Calvo and Keijzer 2009), but not of the biological and comparative psychological research being done to fill it.

1.5 Conclusion

The biological background of the debate over plant cognition involves the examination of more species throughout phylogeny in a more fine-grained manner, motivated and guided by genetics and Evo-Devo. Plants are an important kingdom to explore as we begin to develop behavioral and cognitive characters and continue to improve the genetic, morphological and developmental characters we already have.

Plant advocates show that we can legitimately subsume plants within the cybernetic-computational framework and legitimately use CC concepts to explain their behavior (along with that of the rest of biology) *qua* cybernetic systems. But these are not the best explanations we can provide of their behavior – or of animal behavior, for that matter. I am not ruling out plant cognition in a biologically based framework of cognitive characters, but right now we are far from showing it.

In the meantime, I agree with Chamovitz (2018) that plant science will continue to proceed in exciting directions even without talk of plant intelligence and cognition. However, I disagree that such talk is merely a matter of taste. It can do real damage. First, as we develop a framework of cognitive characters that we can map to phylogeny – potentially including at least some plant species – a motivated cohort of plant advocates appears willing to use cybernetic homonyms that inject unnecessary ambiguity into the scientific discussion when it is not explicitly or obviously metaphorical. Second, some science communicators, including some plant scientists and advocates when writing for the public, will be encouraged to use mindy terms (cognition, intelligence, consciousness, knowledge) that resonate strongly with the public even if their meanings are not necessarily what the public thinks they mean.

I doubt the second problem can be resolved: the professional and financial motivations for using CC concepts widely are substantial. But the first problem can be. Advocates have a choice. They can continue to use cybernetic-computational language for plants in scientific venues while acknowledging that the explanation of plant behavior this provides comes from subsuming it within cybernetics. Alternatively, they can leverage their interest in plants to help reconceptualize plant behavioral flexibility within the phylogenetic context in which plant biologists work. This would enable them to contribute to the development of cognitive characters, thought it would mean not prejudging which, if any, cognitive characters plants might have.

Acknowledgments

I wish to thank Gabriele F., Peter S. and Markus W. for bringing together the Basel plant cognition workshop that led to this volume and sparked many great discussions about the topic. I also thank them and an anonymous reviewer for very helpful feedback that motivated key improvements to my chapter.

Notes

1 Some consider IBE a distinct argument form rather than a subtype of analogy, but this difference, if any, won't matter here.
2 For example, Baluška and Levin (2016), following Lyon (2015), define cognition in terms of information processing – "the total set of mechanisms that underlie information acquisition, storage, processing, and use, at any level of organization"; then, in their section on plant cognition, write that plants "are still considered generally outside of neuronal and cognitive organisms" but "have many features that are considered neuronal"; then, after surveying mechanisms, declare "Plants are emerging as excellent computational systems". Calvo (2016, 1324–5, 1334–5) emphasizes the flexibility of response to environmental contingencies (a.k.a. "non-hardwired strategies") that justifies the ascription of intelligence (a.k.a. "cognitivist information-processing") to plants and correctly notes this is compatible with anti-representationalist views. Calvo and Keijzer (2011, 156) move from "flexible" behavior to "unquestionably intelligent" behavior.
3 Tompkins and Bird may not have intended anything they ascribed to plants metaphorically; any non-metaphorical uses however can be assimilated into the argument from analogy.
4 I should add that the other papers, besides Cvrckova et al. (2016), cited by Segundo-Ortin and Calvo (ibid.) in support of the IBE – Karban (2008), Silvertown and Gordon (1989) – also do not provide this support. They *do* focus on plant growth as the primary means by which plants behave and note the flexibility in their growth – i.e. their developmental plasticity, in Evo-Devo terms. But they do not draw the conclusion advocates do from these facts.
5 The term "analogy" in biology is also used to mean non-homology (homoplasy, convergent or independent evolution). This is distinct from the similarity-based

concept at work in arguments from analogy. For example, Barron and Klein (2016; cited approvingly by Segundo-Ortin and Calvo 2023) argue from analogy that honeybees might be conscious because they share some neural traits with organisms that by wide consensus do have conscious experience. A separate issue is whether honeybees and mammals share the same consciousness character, and if so whether this is by homology or homoplasy (analogy in the non-homology sense).

6 An alternative label might be Allen's (2017) no-definition definition of cognition as "adaptive information-processing". However, the Darwinian flavor of "adaptive" is misleading; this conceptual framework is not restricted to evolved information-processing systems. My emphasis on its cybernetic roots makes this important point clear. See also endnote 8.

7 An important problem in animal cognition is how to untangle cognitive explanations of animals that draw on distinct cognitive-conceptual schemes: anthropocentric, phylogenetic and cybernetic. I set aside this problem here.

8 Thus, when advocates confidently assert there is "no principled reason to deny that radically different neural structures could give rise to felt states" (Segundo-Ortin and Calvo 2023, 2), it is mysterious what principled reason they have in mind other than the medium-independence of the CC framework.

9 This is a very first approximation. For an initial introduction, see Figdor (2022) and the many citations there.

10 Miguel-Tomé and Llinás (ibid., 9) also toss out the possibility of defining an even more general concept of a nervous system that would include AI systems "inspired by" biological organisms. What this concept would pick out is anyone's guess. The biological one is well-defined, however; in their terms, it is not "phylogenetic" in the sense that it is not specific to the animal kingdom because it does not require neurons (but it does require multicellularity).

References

Abramson, Charles I., and Ana M. Chicas-Mosier. 2016. "Learning in Plants: Lessons from Mimosa pudica." *Frontiers in Psychology* 7. https://doi.org/10.3389/fpsyg.2016.00417

Adamatzky, Jordi Vallverdu, Antoni Gandia, Alessandro Chiolerio, Oscar Castro, and Gordana Dodig-Crnkovic. 2023. "Fungal Minds." In *Fungal Machines: Sensing and Computing with Fungi,* edited by Andrew Adamatzky, 409-422. Cham: Springer. https://doi.org/10.1007/978-3-031-38336-6_26.

Adams, Fred. 2018. "Cognition Wars." *Studies in History and Philosophy of Science Part A* 68: 20–30. https://doi.org/10.1016/j.shpsa.2017.11.007

Allan, Colin. 2017. "On (not) Defining Cognition." *Synthese* 194 (3): 1–17. https://link.springer.com/article/10.1007/s11229-017-1454-4.

Alpi, Amedeo, Nikolaus Amrhein, Adam Bertl, Michael R. Blatt, Eduardo Blumwald, Felice Cervone, Jack Dainty, et al. 2007. "Plant Neurobiology: No Brain, no Gain?" *Trends in Plant Science* 12 (4): 135–6. https://doi.org/10.1016/j.tplants.2007.03.002.

Baluška, František, and Michael Levin. 2016. "On Having No Head: Cognition Throughout Biological Systems." *Frontiers in Psychology* 7: 902. https://doi.org/10.3389/fpsyg.2016.00902.

Baluška, František, and Stefano Mancuso. 2007. "Plant Neurobiology as a Paradigm Shift Not Only in the Plant Sciences." Plant Signaling & Behavior 2 (4): 205–7. https://doi.org/10.4161/psb.2.4.4550.

Baluška, František, and Stefano Mancuso. 2009. "Plants and Animals: Convergent Evolution in Action?" In *Plant-environment Interactions: From Sensory Plant Biology to Active Plant Behavior*, edited by František Baluška, 285-301. Berlin, Heidelberg: Springer. https://doi.org/10.1007/978-3-540-89230-4_15.

Baluška, František, Dieter Volkmann, and Stefano Mancuso, eds. 2006. *Communication in Plants: Neuronal Aspects of Plant Life*. Berlin, Heidelberg: Springer. https://doi.org/10.1007/978-3-540-28516-8_2.

Bassel, George W. 2018. "Information Processing and Distributed Computation in Plant Organs." *Trends in Plant Science* 23 (11): 994–1005. https://doi.org/10.1016/j.tplants.2018.08.006.

Bennett, Maxwell R., and Peter M. S. Hacker. 2003. *Philosophical -Foundations of Neuroscience*. ;Oxford ;(UK): Blackwell Publishing.

Birch, Jonathan. 2023. "Disentangling sentience from developmental plasticity." *Animal Sentience* 33 (20): 1-3.

Bose, Jagadish Chandra. 1926. *The Nervous Mechanism of Plants*. London: Longmans, Green, and Co.

Burghardt, Gordon, and Robert I. Bowers. 2017. "From Instinct to Behavior Systems: An Integrated Approach to Ethological Psychology." In *APA handbook of Comparative Psychology: Basic Concepts, Methods, Neural Substrate, and Behavior*, edited by Josep Call, Gordon M. Burghardt, Irene M. Pepperberg, Charles T. Snowdon, and Thomas Zentall, 333–64. ;Washington, D.C: American Psychological Association.

Burkhardt, Pawel, Jeffrey Colgren, Astrid Medhus, Leonid Nigel, Benjamin Naumann, Joam J. Soto-Angel, and Eva-Lena Nordmann. 2023. "Syncytial Nerve Net in a Ctenophore Adds Insights on the Evolution of Nervous Systems." *Science* 380 (6642): 293–7. https://doi.org/10.1126/science.ade5645.

Calvo, Paco. 2016. "The -Philosophy of Plant Neurobiology: A Manifesto." *Synthese* 193: 1323–43. https://doi.org/10.1007/s11229-016-1040-1.

Calvo, Paco. 2017. "What Is it Like to Be a Plant?" *Journal of Consciousness Studies* 24 (9–10): 205–27.

Calvo, Paco, and Fred Keijzer. 2009. "Cognition in -Plants." In *Plant-environment Interactions: From Sensory Plant Biology to Active Plant Behavior*, edited by František Baluška, 247–66. Berlin, Heidelberg: Springer. https://doi.org/10.1007/978-3-540-89230-4_13.

Calvo, Paco, and Fred Keijzer 2011. "Plants: Adaptive Behavior, Root-brains, and Minimal Cognition." *Adaptive -Behavior* 19 (3): 155–71. https://psycnet.apa.org/doi/10.1177/1059712311409446.

Calvo, Paco, Monica Gagliano, Gustavo M. Souza, and Anthony Trewavas. 2020. "Plants Are Intelligent, Here's How." *Annals of Botany* 125 (1): 11–28. https://doi.org/10.1093/aob/mcz155.

Chamovitz, Daniel. A. 2018. "Plants Are Intelligent; Now What?" *Nature Plants* 4: 622–3. https://doi.org/10.1038/s41477-018-0237-3.

Chen, Zhuo, and John J. Wiens. 2020. "The Origins of Acoustic Communication in Vertebrates." *Nature Communications* 11: 369. https://doi.org/10.1038/s41467-020-14356-3.

Chirimuuta, Mazviita. 2021. "Your Brain is Like a Computer: Function, Analogy, Simplification." In *Neural Mechanisms: New Challenges in the Philosophy of Neuroscience*, edited by Fabrizio Calzavarini and Marco Viola, 235–61. Cham: Springer.

Cvrčková, Fatima, Helena Lipavská, and Viktor Žárský. 2009. "Plant Intelligence. Why, Why Not or Where?" *Plant Signaling & Behavior* 4 (5): 394–9. https://doi.org/10.4161/psb.4.5.8276.

Cvrckova, Fatima, Viktor Zarsky, and Anton Markos. 2016. "Plant studies may lead us to rethink the concept of behavior". Frontiers in Psychology 7: 622. 1-4. *doi: 10.3389/fpsyg.2016.00622*

Darwin, Charles. 1859. *On the Origin of Species*. New York: F.F. Collier and Son.

Darwin, Charles. 1880. *The Power of Movements in Plants*. Edinburgh: John Murray.

Delaux, P. M., Alexander J. Hetherington, Yoan Coudert, Charles Delwiche, Christophe Dunand, Sven Gould, Paul Kenrick, et al. 2019. "Reconstructing Trait Evolution in Plant Evo–devo Studies." *Current Biology* 29 (21): R1110–18. https://doi.org/10.1016/j.cub.2019.09.044.

de Queiroz, Allan, and Peter H. Wimberger. 1993. "The Usefulness of Behavior for Phylogeny Estimation: Levels of Homoplasy in Behavioral and Morphological Characters." *Evolution* 47 (1): 46–60. https://doi.org/10.1111/j.1558-5646.1993.tb01198.x.

Figdor, Carrie. 2018a. *Pieces of Mind: The Proper Domain of Psychological Predicates*. New York: Oxford University Press.

Figdor, Carrie. 2018b. "The Rise of Cognitive Science in the 20th Century." In *Philosophy of Mind in the Twentieth and Twenty-First Centuries: The History of the Philosophy of Mind*, edited by Amy Kind, 280–302. London: Routledge. http://dx.doi.org/10.4324/9780429508127-12.

Figdor, Carrie. 2022. "What Could Cognition Be, if Not Human Cognition? Individuating Cognitive Abilities in the Light of Evolution." *Biology & Philosophy* 37 (6): 52. http://dx.doi.org/10.1007/s10539-022-09880-z.

Figdor, Carrie 2023. "What are we talking about when we talk about cognition?" JoLMA: the Journal for the Philosophy of Language, Mind, and the Arts. 4 (2): 149-162. DOI 10.30687/Jolma/2723-9640/2023/02/001.

Finet, Cédric, and Yvon Jallais. 2012. "Auxology: When Auxin Meets Plant Evo-devo." *Developmental Biology* 396 (1): 19–31. https://doi.org/10.1016/j.ydbio.2012.05.039.

Firn, Richard. 2004. "Plant Intelligence: An Alternative Point of View." *Annals of Botany* 93 (4): 345–51. https://doi.org/10.1093/aob/mch058.

Gagliano, M., Vladyslav Vyazovskiy, V., Alexander A. Borbély, Mavra Grimonprez, and Martial Depczynski. 2016. "Learning by Association in Plants." *Scientific Reports* 6 (1): 38427. https://doi.org/10.1038/srep38427.

Holmes, E., and G. Gruenberg. 1965. "Learning in Plants." *Worm Runner's Dig* 7: 9–12.

Karban, Richard. 2008. "Plant Behaviour and Communication." *Ecology Letters* 11 (7): 727–39. https://doi.org/10.1111/j.1461-0248.2008.01183.x.

Keijzer, Fred. 2013. "The Sphex Story: How the Cognitive Sciences Kept Repeating an Old and Questionable Anecdote." *Philosophical Psychology* 26 (4): 502–19. http://dx.doi.org/10.1080/09515089.2012.690177.

Klein, Colin, and Andrew B. Barron. 2016. "Insects Have the Capacity for Subjective Experience." *Animal Sentience* 1 (9). http://dx.doi.org/10.51291/2377-7478.1113.

Lee, Jonny, Miguel Segundo-Ortin, and Paco Calvo. 2023. "Decision Making in Plants: A Rooted Perspective." *Plants* 12 (9): 1799. https://doi.org/10.3390/plants12091799.

Levin, M., Fred Keijzer, Pamela Lyon, and Detlev Arendt. 2021. "Uncovering Cognitive Similarities and Differences, Conservation and Innovation." *Philosophical Transactions of the Royal Society B* 376 (1821): 20200458. https://doi.org/10.1098/rstb.2020.0458

Lyon, Pamela. 2015. "The Cognitive Cell: Bacterial Behavior Reconsidered." *Frontiers in Microbiology* 6: 264. https://doi.org/10.3389/fmicb.2015.00264.

Lyon, Pamela, Fred Keijzer, Detlev Arendt, and Michael Levin. 2021. "Reframing Cognition: Getting Down to Biological Basics." *Philosophical Transactions of the Royal Society B*, 376 (1820): 20190750. https://doi.org/10.1098/rstb.2019.0750.

Markel, Kasey. 2020. "Lack of Evidence for Associative Learning in Pea Plants." *ELife* 9: e57614. https://doi.org/10.7554/eLife.57614.

Mediano, Pedro A.M., Anthony Trewavas, and Paco Calvo. 2021. "Information and Integration in Plants. Towards a Quantitative Search for Plant Sentience." *Journal of Consciousness Studies* 28 (1–2): 80–105.

Meyerowitz, E. M. 2002. "Plants Compared to Animals: The Broadest Comparative Study of Development." *Science* 295 (5559): 1482–5. https://doi.org/10.1126/science.1066609.

Miguel–Tomé, Sergio, and Rodolfo R. Llinás. 2021. "Broadening the -Definition of a Nervous System to Better Understand the Evolution of Plants and Animals." *Plant Signaling & Behavior* 16 (10): 1927562. https://doi.org/10.1080/15592324.2021.1927562.

Miłkowski, Marcin. 2018. "From Computer Metaphor to Computational Modeling: The Evolution of Computationalism." *Minds and Machines* 28: 515–41. https://doi.org/10.1007/s11023-018-9468-3.

Money, Nicholas P. 2021. "Hyphal and Mycelial Consciousness: The Concept of the Fungal Mind." *Fungal Biology* 125 (4): 257–9. https://doi.org/10.1016/j.funbio.2021.02.001.

Moroz, Leonid L., and Daria Y. Romanova. 2021. "Selective Advantages of Synapses in Evolution." *Frontiers in Cell and Developmental Biology* 9: 726563. https://doi.org/10.3389/fcell.2021.726563.

Moroz, Leonid L. and Daria Y. Romanova. 2022. "Alternative neural systems: What is a neuron? (Ctenophores, sponges, and placozoa). Frontiers in Cell and Developmental Biology 1-19. *doi: 10.3389/fcell.2022.1071961.*

Olson, Mark E. 2019. "Plant Evolutionary Ecology in the Age of the Extended Evolutionary Synthesis." *Integrative and Comparative Biology* 59 (3): 493–502. https://doi.org/10.1093/icb/icz042.

Rendall, Drew, and Anthony Di Fiore. 2007. "Homoplasy, Homology, and the Perceived Special Status of Behavior in Evolution." *Journal of Human Evolution* 52 (5): 504–21. https://doi.org/10.1016/j.jhevol.2006.11.014.

Segundo-Ortin, Miguel, and Paco Calvo. 2019. "Are Plants Cognitive? A Reply to Adams." *Studies in History and Philosophy of Science Part A* 73: 64–71. https://doi.org/10.1016/j.shpsa.2018.12.001.

Segundo-Ortin, Miguel, and Paco Calvo. 2022. "Consciousness and -Cognition in Plants." *Wiley Interdisciplinary Reviews: Cognitive Science* 13 (2): e1578. https://doi.org/10.1002/wcs.1578.

Segundo-Ortin, Miguel, and Paco Calvo. 2023. "Plant Sentience? Between Romanticism and Denial: Science." *Animal Sentience* 33 (1). http://dx.doi.org/10.51291/2377-7478.1772.

Shapiro, James A. 2021. "All Living Cells Are Cognitive." *Biochemical and Biophysical Research Communications* 564: 134–49. https://doi.org/10.1016/j.bbrc.2020.08.120.

Silvertown, Jonathan, and Deborah M. Gordon. 1989. "A Framework for Plant Behavior." *Annual Review of Ecology and Systematics* 20: 349–366. https://doi.org/10.1146/annurev.es.20.110189.002025.

ten Cate, Carel. 2023. "Plant Sentience: A Hypothesis Based on Shaky Premises." *Animal Sentience* 33 (13). http://dx.doi.org/10.51291/2377-7478.1795.

Tompkins, Peter, and Cristopher Bird. 1974. *The Secret Life of Plants.* London: Allen Lane.

Trewavas, Anthony. 2003. "Aspects of Plant Intelligence." *Annals of Botany* 92 (1) 1–20. https://doi.org/10.1093/aob/mcg101.

Trewavas, Anthony. 2005. "Plant Intelligence." *Die Naturwissenschaften* 92 (9): 401–13. https://doi.org/10.1007/s00114-005-0014-9.

Trewavas Anthony. 2007. "Response to Alpi et al.: Plant Neurobiology--all Metaphors Have Value." *Trend in Plant Science* 12 (6): 231–33. https://doi.org/10.1016/j.tplants.2007.04.006.

Trewavas, Anthony. 2016. "Plant Intelligence: An Overview." *BioScience* 66 (7): 542–51. https://doi.org/10.1093/biosci/biw048.

Wiener, Norbert. 1948/2019. *Cybernetics or Control and Communication in the Animal and the Machine.* Cambridge, MA: MIT Press.

Woese, C. R., L. J. Magrum, and G. E. Fox. 1978. "Archaebacteria." *Journal of Molecular Evolution* 11: 245–52. https://doi.org/10.1007/BF01734485.

2 What Deception Reveals about Plant Cognition

Marc Artiga

2.1 Introduction

The goal of this chapter is to provide an original argument for the view that many plants (perhaps all) should be considered cognitive systems.[1] Let's call such a view "plant cognitivism" (PC):

PLANT COGNITIVISM (PC): Many plants are cognitive systems.

PC can be defended following what we might call a "direct" or an "indirect" strategy. A *direct* strategy aims to show that plant's adaptive behaviour is best explained by attributing to them those capacities typically studied by cognitive science, such as knowledge, memory, learning, decision-making or anticipation. Most researchers interested in PC follow this path (Calvo 2011; Calvo et al. 2020; Trewavas 2003, 2009; Gagliano 2017). However, a recurring difficulty with this strategy is that it requires a non-controversial definition of knowledge, memory and the like. A narrow definition of, say, "knowledge" undermines plant cognitivism, whereas a broader definition leaves room for it. Thus, the debate tends to turn on whether a narrow or a broader definition is preferable. In that respect, adopting a *piecemeal* approach (Lee 2023) that addresses particular capacities, rather than "cognition" as such, does not assuage this worry.

Instead of pursuing a direct strategy, here I will follow an *indirect* one: rather than concentrating on what plants are capable of doing, I will focus on what can be done to them. The way an organism responds to a certain situation can reveal important aspects of its inner life. Thus, in what follows, I will argue that a particular phenomenon that takes place between organisms provides evidence for plant cognitivism. In other words, the existence of a certain pattern of interaction and the plant's reaction to it lends support to the idea that plants are cognitive systems. By moving to an indirect strategy, we should be able to avoid (or minimize) the worry that we are engaging in a mere terminological quibble.[2]

DOI: 10.4324/9781003393375-4

More precisely, I will discuss communication and deception and I will defend two claims, one negative and one positive. First, in contrast to some of the views expressed in the literature, the existence of communication or deception by themselves fails to provide evidence for plant cognitivism. Nonetheless, I will also argue that there is a specific combination of these elements that *does* provide evidence for the cognitive status of plants; certain kinds of interactions that involve communication and deception (but not others) suggest that plants have at least a minimal form of cognition.

In a nutshell, the main argument of the positive part of chapter runs as follows. Plants can be deceived. This is not just a conceptual point; I will present some recent evidence suggesting that plants are actually fooled by certain sorts of signals. In addition, I will argue that only organisms that have some minimal form of cognition can be deceived – only cognitive systems can be fooled. As a result, the vegetal capacity of being deceived reveals their possession of some minimal form of cognition.

The chapter is structured as follows. Section 2.2 develops the negative side of the argument: I will argue that not just any form of communication or deception supports plant cognitivism. Standard models suggest that from the mere existence of communication one cannot infer anything about the cognitive capacities of the sender or the receiver. Similarly, the fact that an organism can deceive others fails to reveal any aspect of its cognitive endowment.

Section 2.3 begins to develop the positive side: I will identify some specific aspects of deception that *do* provide evidence for PC. More precisely, I will argue that, even though deceivers need not possess any specific capacities (such as memory or conceptualization), only cognitive systems can be deceived. In other words, deception requires a certain minimal form of cognition on the part of the receiver.

In Section 2.4 I will present some recent empirical research suggesting that tomato plants are actually deceived through communicative signals. Deception taking place via communication will play a crucial role in my response to some important objections. Finally, I will argue that the fact that tomato plants can be deceived lends support for plant cognitivism (PC). In Section 2.5 I will summarize the main results and discuss some future lines of research.

2.2 The Negative

Communication and deception are some of the most complex forms of human interaction. Accordingly, one might plausibly assume that if plants are shown to communicate and deceive, one could straightforwardly build an argument for plant cognitivism. In this first part of the chapter, I will examine and criticize this intuitive reasoning: in spite of some claims that we can find in the literature, not any form of communication or deception in plants provides evidence for PC. The main goal of this section is to show that the mere existence of communication or deception in plants falls short of vindicating PC.

2.2.1 Communication

It is a well-established fact that plants communicate not only between each other but also with other organisms such as animals and bacteria (Baluška, Volkmann, and Mancuso 2006; Baluška and Mancuso 2009; Witzany et al. 2012). Plant communication is a complex phenomenon: for one thing, it is carried out by different means such as airborne signalling, hormones, electrical signalling or chemical communication in the rhizosphere. In turn, each of these strategies raises its own challenges: for instance, the blends employed in airborne communication or their specific receptors are still not well understood. Furthermore, signalling causes many different kinds of behaviours on the receiver's side. Partly for these reasons, plant communication is a growing area of research.

Since plants are sessile organisms, what is the function of plant communication? Consider airborne signalling (Farmer 2001; Kessler et al. 2023). A prominent function of this form of communication is to provide a defence strategy against pests. Wounded plants emit volatile organic compounds (VOCs) that induce various defence mechanisms in undamaged plants. This phenomenon is called "priming", which can actually be regarded as a form of behavioural plasticity (Agrawal 1999). Plants that receive VOCs, for instance, might be more attractive to parasitoids, more resistant to pests or exhibit a higher response to wounds. Airborne signals are used to prime other plants but also other parts of the self-same plant; since different branches might be nearby, but distant in branching structure, VOCs are often the most rapid, precise and efficient way of sending information between different parts of the same organism. Thus, VOCs are classified as signals partly because assuming that they carry information about predators, etc. helps to explain behaviour.

Now, is the capacity to send signals a mark of cognition? Certainly, in general some forms of communication are closely connected to certain cognitive capacities; Gricean communication, for instance, might require specific mental capacities, like complex intentions (Tomasello 2008; Scott-Phillips 2014). However, the existence of communication as such is not necessarily connected to any specific cognitive ability (cf. Calvo and Keijzer, 2011; Gross 2016). On the standard definition, a signal is "any act or structure that alters the behaviour of other organisms, which evolved because of that effect, and which is effective because the receiver's response has also evolved" (Maynard-Smith and Harper 2004, 3). This is fully compatible with two living beings engaging in some form of communication without having an inner cognitive life. For instance, microbiologists accept that bacteria communicate, even if many would resist the idea of attributing them cognitive capacities (Miller and Bassler 2001; Papenfort and Bassler 2016; Artiga 2021). Indeed, it has even been suggested that viruses might engage in some form of communication (Dolgin 2019). The fact that two organisms communicate via signals tells us something important about their interaction, but not necessarily about their cognitive endowment. This is even true if, as a matter of fact, most species that have the capacity to send signals qualify as cognitive systems. As such, communication between two individuals does not seem to be a mark of the cognitive. In the same way that the complexity of a signalling molecule does not imply a complex meaning (Firn, 2004), *prima facie* the complexity of meaning does not seem to entail sophisticated processing either on the part of the sender or the receiver.[3]

Although the connection between interorganismic communication and cognition is unclear, the link between intraorganismic signalling and cognition might be tighter. In particular, it might be that communication within an individual is a requirement for intelligence. If intelligent behaviour involves combining information in flexible ways, the use of memory in the right circumstances, or taking into account different bits of relevant information to make a decision, then cognition requires some form of communication or signalling between different parts of an organism (Trewavas 2003; Bassel 2018; Godfrey-Smith 2020). Nonetheless, two features of this idea are worth emphasizing. First, intraorganismic communication is obviously not sufficient for intelligent behaviour, since it also crucially depends on how this information is put to use (e.g. Firn 2004). Accordingly, to vindicate PC it has to be shown that intraorganismic signalling constitutes a genuine form of communication between different parts of the same individual that can ground the emergence of cognitive capacities such as memory, decision-making or learning. Second, note that arguing in favour of this claim actually amounts to pursuing a *direct* strategy for PC, as one is assessing whether concepts like "learning" or "decision-making" apply to plants by focusing on what these organisms can do. Since in this chapter I want to follow a different, *indirect* approach, I will leave aside the question of the relationship between PC and intraorganismic communication.

To sum up, whereas the existence of plant-plant, plant-animal and plant-bacteria communication is an important and valuable discovery, it is very hard to build a case for PC solely on that base. The existence of communication as such fails to provide evidence for plant cognitivism. As will be seen in Section 2.4, however, it might constitute an essential ingredient of a compelling argument.

2.2.2 Deception

Many flowering plants (angiosperms) use petal colour patterns to indicate the presence of nectar and lure insects. Others mimic these patterns to attract pollinators without providing any reward. This is typically described as an instance of deception. Some plants go even further and mimic animals; famously, orchids in the genus *Ophrys* resemble female bees or wasps to attract males, which attempt to copulate with them and end up spreading the plant's pollen (Schiestl 2005; Anders Nilsson 1992). Analogously, a flower from the genus *Aristolochia* uses smell of rotting flesh to lure flies (Oelschlägel et al. 2015). Although the available evidence for plant mimicry was questioned some years ago (Roy and Widmer 1999), subsequent experiments confirmed the existence of plant deception (Benitez-Vieyra et al. 2007; Peter and Johnson 2008).

Two aspects of this research are worth noting. First, most cases of plant deception (perhaps all) involve mimicry. For instance, orchids mimic female wasps and *Aristolochia* flowers mimic rotting flesh. The second important point is that research on plant deception almost invariably addresses cases of plants *deceiving* animals, rather than plants *being deceived*. Again, orchids and *Aristolochia* flowers provide beautiful illustrations.

Now, can the existence of plant deception vindicate plant cognitivism? Again, probably not. The fact that plants can deceive does not indicate that they have a minimal form of cognition (cf. Calvo and Keijzer 2011, 159). An organism need not be smart to fool another. Anything that evolves can deceive by acquiring a look that mimics the right kind of model. Drone flies, stick insects or common seadragons need not be intelligent to mislead: they just need to well enough resemble bees, branches and seaweed, respectively. The fact that orchids lure male wasps with their appearance can hardly tell us anything interesting about their cognitive endowment. Thus, an argument for PC cannot solely be based on plant deception.

In conclusion, *prima facie* the mere existence of plant communication or deception does not provide evidence for plant cognition. I think this negative result is of some relevance, since these striking phenomena are often mentioned in arguments for PC. For example, in his review Gross (2016) claims that communication and pollinator deception are two of the main capacities suggesting that plants are cognitive systems. If my arguments so far are on the right track, these phenomena are certainly remarkable in many respects, but silent as regards PC. More work needs to be done to develop a convincing argument for plant cognitivism.

2.3 The Positive

I have shown that neither communication nor deception by themselves provide evidence for plant cognitivism. In this section, however, I will argue that a powerful argument can be built upon these elements.

Let us consider deception again in more detail and, in particular, let us focus on deception based on mimicry, which is the relevant category for our purposes. Think of three paradigmatic examples: drone flies, which resemble bees and thus avoid being preyed upon; *Photuris* fireflies, which attract males and devour them by imitating the flashings of female fireflies of other species; and stick insects. What is common between these cases? First, deception requires a sender (the *deceiver*) and a receiver (the *deceived*).[4] Second, the sender produces something (e.g. some skin colour, odour, shape, behaviour) that resembles a model in certain respects. Deception generally works by pretending to be something else. These two features, however, are found in many mechanisms that do not involve deception, so something important seems to be missing. Consider, for instance, Müllerian mimicry: Monarch and Viceroy butterflies resemble each other and both are equally poisonous. Apparently, both benefit from sharing an advertisement display, because predators learn faster to avoid them: since predators cannot distinguish them, after eating an individual of one of the two species, predators will learn to avoid both of them. In the framework we have been developing, Viceroy and Monarchs are senders that develop similar patterns of wing colours, which are perceived by the same receivers, but are not deceiving them. Rather, this is usually interpreted as *hiding* truth: individuals do not reveal which one of the two species they belong to.

So what distinguishes examples of mimicry in which individuals hide truth from those in which there is genuine deception? Here is a plausible suggestion: in deception the mimic not only resembles the model in some respect, but the function of this similarity is to cause the receiver to *misrepresent* the world in certain respects (Artiga and Paternotte, 2018; forthcoming). Drone flies (*Eristalis tenax*), for instance, mimic the bee's colour and behaviour to cause predators to miscategorize them as bees. *Photuris* fireflies mimic the flashings of female fireflies willing to mate and are (wrongly) classified as such; when males approximate, *Photuris* fireflies devour them. Stick insects resemble branches and are misclassified as branches.

"Misrepresentation" should be understood very broadly here; to misrepresent is simply to have a state with false representational content. Standard cases of misrepresentation caused by deceptive mimicry typically involve miscategorization or misperception, but other forms of misrepresentation that do not fit these categories are also possible. In this sense, the capacity to misrepresent does not presuppose the possession of compositional thought, consciousness or concepts (even if, as some argue, categorization requires concept possession; for a discussion, see Mandelbaum 2017; Millikan 2017; Deroy 2019). Similarly, I assume that truth and falsity can be attributed to non-linguistic representations, which is a common presupposition in the literature (Millikan 1984; Neander 2017; Shea 2018). An alarm call performed by a ground squirrel, for instance, can be true or false depending on whether a predator is actually approaching; likewise, truth and falsity also apply to the squirrel's mental state.

My claim, then, is that deception (or, at least, deception via mimicry, which, for our purposes, is the important phenomenon) necessarily involves misrepresentation (typically miscategorization or misperception) on the part of the receiver. This should be interpreted as a necessary condition, rather than as a definition. I am not here so much interested in providing necessary and sufficient conditions for deception to occur (which would require a paper of its own) but to identify a feature that accompanies all cases. This is all I need for the argument to succeed.

Let me provide three considerations that lend support to the idea that receiver misrepresentation is a necessary condition for deceptive mimicry. First, scientific research on this topic requires assessing, not how much two organisms seem similar to us, but whether they look similar to a specific consumer. Mimicry is in the eye of the beholder (Font 2019). As Shaefer and Ruxton claim (2009, 681), "the fitness landscapes of mimics and of species that exploit perceptual biases are both a product of the perception of the beholder". Adopting this perspective has turned out to be crucial: as a matter of fact, it turns out that some organisms that look very similar to us can actually be distinguished by the relevant actors (e.g. using ultraviolet vision) and others that seem obviously different are undistinguishable to them. Accordingly, convincing evidence for deceptive mimicry has to show that mimic and model resemble each other well enough from the point of view of the consumer, which is something that requires assessing the receiver's cognitive capacities (with special focus on perception and categorization).

Second, consider the following question: why is plant mimicry (and deception) relatively uncommon as compared to animal mimicry? One plausible explanation suggests that it is difficult for pollinators to learn which plants are rewarding and which are not, due to several mechanisms: because of low frequency of unrewarding plants, because it is difficult to distinguish unrewarding plants from those that have recently been depleted by another pollinator, etc. In all these explanations, the cognitive capabilities (learning, memory, etc.) of receivers partly determine the evolution of mimicry and deception (Shaefer and Ruxton 2009, 682). Again, the receivers' cognitive capacities play a central role in understanding the evolution of mimicry and deception.

Finally, a conceptual motivation for adding the misrepresentation requirement can also be defended. A moment's reflection suggests that, intuitively, there has to be something minimally cognitive in being deceived; being deceived is not just being caused to act in a certain way; it has something to do with causing you to behave in a certain way because someone led you to maintain or acquire a false perspective on the world. Deception shapes behaviour by causing misinformation. Among other things, it is an epistemic harm, and that requires some form of misrepresentation, broadly understood. Consequently, in what follows I will assume that only organisms that can misrepresent the world (typically, miscategorize or misperceive it) can be deceived. An organism cannot be fooled unless it can possess false representations of the world. This will be a key premise of the main argument of the chapter.

I previously argued that the existence of plant deception does not imply any sophisticated cognitive capacity in either sender or receiver. I maintained this because, typically, when scientists talk about plant deception they refer to situations in which plants play the role of deceiver. *Prima facie*, deceivers need not be intelligent or possess any specific cognitive capacity. I have then argued, however, that deceived organisms must be able to misrepresent the world. In particular, deceived organisms typically miscategorize or misperceive some aspect of the world. Thus, if we find evidence of plants being deceived, that will in turn provide evidence for some form of misrepresentation, which is, essentially, a minimal form of cognition.

To be as clear as possible, let me pause a second and clarify the connection between deception, misrepresentation and cognition. I argued that an organism can be deceived only if it is able to misrepresent the world as being a certain way. Typical forms of misrepresentation involved in deception comprise misperception or miscategorization of a certain entity. In turn, these capacities imply that organisms are able to represent (typically perceive and categorize) the world as being a certain way. Finally, I consider these representational capacities constitute minimal forms of cognition. In other words, each of the following concepts constitutes a sufficient condition for the next: being deceived, being able to misrepresent, being able to represent, being a cognitive system.

Accordingly, the question to be addressed is the following: can plants be deceived? In what follows I will defend an affirmative answer. In particular, in the following section I will present a case in which they are actually deceived. I will also defend my interpretation and draw some consequences from it.

2.4 Deceived Plants

As stated above, in an overwhelming number of cases research on plant deception investigates how plants are able to deceive other organisms, rather than discussing whether plants can be deceived. Nonetheless, within this vast literature, I would like to discuss one case (or, perhaps, a category of cases) that differs from the rest. In particular, I will argue that, on the most plausible interpretation, this is an example of *plants being deceived*.

Interestingly, the example combines deception and communication in a way that is relevant for my argument.

As explained in Section 2.1, plants exhibit a variety of responses to attack depending on the kind of danger. Two particularly common defences depend on the salicylic acid (SA) signalling pathway, which is most effective against biotrophic organisms (typically pathogens), and the jasmonic acid (JA) signalling pathway, which prepares the plant against necrotrophic agents (typically some insect). Jasmonic acid, for instance, induces antibiosis, antixenosis and recruitment of natural enemies (Stout and Duffy 1996; Cooper and Goggin 2005). Furthermore, it triggers an increase in the density of glandular trichomes on the leaf surface (Traw and Bergelson 2003; Boughton, Hoover, and Felton 2005). Trichomes produce various sorts of sticky and/or toxic exudates that irritate, entrap or poison herbivores (Simmons and Gurr 2004). Interestingly, a trade-off exists between different defence strategies, since there are certain costs associated with them (Agrawal 1999; Agrawal et al. 2002; Traw and Bergelson 2003; Heil and Baldwin 2002). More precisely, there is negative cross-talk between JA- and SA-induced responses (Glazebrook 2005; Thaler, Humphrey, and Whiteman 2012). A SA-regulated defence, for instance, *decreases* the number of density of glandular and non-glandular trichomes (Traw and Bergelson 2003) and triggers cell death (which is effective against biotrophic threats). Likewise, a JA-regulated defence tends to inhibit SA defences.

Now, the hemi-biotrophic pathogen *Pseudomonas syringae* has evolved a remarkable strategy to take advantage of this trade-off. This gram-negative bacterium releases coronatine to activate jasmonic acid-derived responses and block the activation of salicylic acid-mediated responses. In other words, some substance released by *P. syringae* induces plants to develop a maladaptive response, developing JA-associated responses, which work best against insects and, as a result, inhibiting their SA-induced defences, which work best against pathogens like *P. syringae*.

Is this an example of plant being deceived? It is hard to tell. One might argue that the plant is not misrepresenting the environment; instead, the situation could be best described as the insect directly and actively suppressing the plant's defences. Since *P. syringae* directly intervenes on the plant, any effect on the plant's response could be interpreted as some form of non-cognitive manipulation. It certainly causes the plant to behave in a certain (maladaptive) way, but it is unclear if the plant behaves in such a way because it has the wrong kind of information. *P. syringae* might be forcing the plant to behave in a maladaptive way, rather than misleading it.

An analogy may be useful here: anabolic pills cause a bodily change (muscle building), but there is no reason to think this effect is cognitively mediated. Of course, if one already thinks that the human body contains a high number of distributed representations, then the idea that anabolic pills work by altering certain embodied cognitive processes might seem plausible. However, if one is not already convinced of the existence of these internal representations, the effect of anabolic pills is unlikely to help resolve this issue. Likewise, plant cognitivists might happily accept that *P. syringae* deceives the host plant, whereas critics might interpret that the bacterium forces the plant to develop certain defences by exploiting a non-cognitive process. As a result, there are different ways of interpreting the situation and, unless one provides additional considerations, it is hard to know which of them is preferable.

Fortunately, there is a related but different example that may allow us to sidestep these issues. It does not involve a bacterium but an insect: the whitefly *Bemisia tabaci*.

2.4.1 On Whiteflies and Plants

Bemisia tabaci is an hemipteran that feeds by tapping into the phloem of plants, which produces significant damage to the host. In some plants, whitefly attacks trigger JA responses, which work best against insects. However, in others such as the tomato plant, JA levels decline within days, whereas SA responses gradually increase (Zhang et al. 2019). In other words, whiteflies induce infested plants to generate SA defences, which work best against pathogens, instead of JA responses, which in principle are directed at insects like *B. tabaci*. Accordingly, since resources are devoted to the wrong kind of defence, it makes the plant more suitable for whitefly development. In this sense, whiteflies are the mirror image of *P. syringae*: each of these species triggers the response that works best against the other.

Again, the question of whether this is an example of deception is controversial. Walling (2008, 862), for instance, suggests that "[Silverleaf whiteflies] deceive Arabidopsis plants and prevent the activation of the JA-regulated defences that actively deter nymph development". Nonetheless, a critic might object that the plant is not misrepresenting the environment; instead, the situation could best be described as the insect directly and actively suppressing the plant's defences. In other words, since whiteflies pierce the plant, insert toxic saliva and directly intervene on the plant, any effect on the plant's response could be interpreted non-cognitively, as occurred with *P. syringae*.

However, there is a twist in the case of the whitefly: it has been shown that *B. tabaci* not only cause plants to produce the wrong responses, but they also induce them to send volatile organic compounds (VOCs) that cause other plants to produce the wrong responses. VOCs sent by infested plants prime other plants to prepare for a pathogen attack and reduce their defences against whiteflies. As Zhang et al. (2019, 3) suggest:

Whiteflies manipulate future host plants by triggering their host plants to emit a blend of volatiles that signals the deceptive information to neighboring plants.

I think the VOC-mediated interaction provides a much stronger case for an interpretation in terms of deception. Whereas the behaviour of *P. syringae* could be interpreted as a direct manipulation of the host without deception, this is much less plausible in the case of whiteflies causing distant plants to develop the wrong defences via VOCs. Recall that most people accept that VOCs are genuine signals, so they carry certain information, content or meaning. Furthermore, appealing to the information carried by the signal is crucial to explain the receiver's behaviour. Accordingly, if one accepts that plants engage in communication, then the whiteflies' manipulation of plants via VOCs should be interpreted as manipulation by means of informational signals, rather than directly forcing the plant to do something.[5] The most plausible interpretation is that plants behave maladaptively because they acquire the wrong kind of information and, as a result, end up misrepresenting the type of danger. Hence, whiteflies cause plants to acquire the false representation that a biotrophic organism (e.g. a pathogen) is likely to attack them.

Against this analysis, one could object that this description over-intellectualizes plants: a simpler deflationist explanation would suggest that infested plants simply trigger maladaptive behaviour in other plants, but nothing like false representational content need be involved. I think an analogy shows why this reasoning probably is an *ad hoc* move. As the vast literature on plant deception illustrates, it is widely accepted that plants deceive insects by mimicking colour patterns, odour, shape, etc. Now, if plants can deceive pollinators by sending certain molecules, why can plants not deceive other plants in the very same way? I doubt there is any significant difference between some instances of plant-insect and plant-plant deception. For example, in both cases signalling molecules are used that cause maladaptive behaviour (e.g. compare plants that mimic sexual pheromones with plants sending VOCs). Likewise, in all of them deception often triggers a fairly automatic response in the receiver. Those willing to resist the idea that plants are deceived by causing them to misrepresent the world as being certain way need to find a significant difference between the plant-pollinator interaction, which is widely regarded as an instance of genuine deception, and the plant-plant interaction I describe here. Of course, replying that only plant-insect behaviour qualifies as genuine deception because insects are cognitive systems whereas plants are not simply begs the question at stake. In any case, I think the burden of proof is on the opponent: she has to identify a significant difference between plant-insect and plant-plant interactions that can underpin a difference between genuine deception on the one hand and a mere appearance of deception on the other.[6]

In conclusion, I suggest that, on the most plausible interpretation, tomato plants are deceived. This is one of the key premises of my argument for plant cognitivism.

2.4.2 The Main Argument

In previous subsections I put forward and justified the key claims of my argument. Crucially, I argued that some plants are deceived and that, in general, only organisms that can misrepresent the world can be deceived. Although by now the structure of the argument should be obvious, let me present it as clearly as possible. Here is the main argument of the chapter:

> P1 Tomato plants are deceived
> P2 An organism O is deceived only if O misrepresents (typically miscategorizes or misperceives) some aspect of the world
> P3 O misrepresents (typically miscategorizes or misperceives) some aspect of the world only if O is a cognitive system.
> P4 Tomato plants are cognitive systems (from P1 to P3)
> P5 If tomato plants are cognitive systems, many other plants are cognitive systems as well.
> C Therefore, many plants are cognitive systems (Plant Cognitivism)

P2 and P3 have been defended in Section 2.3 and P1 was defended in Section 2.4.1. I would suggest P5 is extremely plausible: there is nothing special about tomato plants. I have focused on tomato plants only because I found a particularly interesting case of deception involving them, but whatever process within tomato plants grounds their ability to misrepresent the world, it is probably shared with many other plants. In that respect, I see no reason for tomato plant exceptionalism. Thus, I think I have justified all premises of the argument and its conclusion vindicates plant cognitivism: many plants are cognitive systems.

Finally, I would like to assuage anti-representationalist worries. Although I laid out my argument in representational terms, I do not think this is essential. The key claim required for the argument to go through is that only organisms that possess certain cognitive capacities (related to categorization, perception and the like) can be deceived. If one's preferred theory explains cognitive phenomena in non-representational terms (e.g. in terms of a complex pattern of enactive engagement with the world, functional misinformation and the like), I think the argument could be simply rephrased using these alternative expressions. As soon as one maintains the connection between minimal forms of cognition and the receiver requirements for deception, the main argument still holds.

2.5 Concluding Remarks

The central goal of the chapter is to provide an original argument in favour of plant cognitivism, i.e. the view that many plants are cognitive systems. The key claims required to build the main argument rely on two ideas: plants can be deceived and only cognitive systems can be deceived. Although both ideas are controversial, I think they are backed up by compelling arguments, or so I have argued.

What are the consequences of adopting this view? I think significant consequences ensue, not only about our understanding of plants but also for our conception of deception. Let me briefly highlight some of them. Two concern our understanding of plants, and the other two of deception.

First, if the main argument is sound, plants should typically be regarded as cognitive systems. A number of interesting consequences follow from this perspective. To name just a few, it would contribute to a better understanding of plant behaviour, direct future research to particular questions that need experimental investigation, lead to better appreciation of evolutionary processes that lead to cognition and intelligent behaviour and, in general, provide a better understanding of what plants really are (Calvo et al. 2020).

Second, in order to establish that plants are cognitive systems, I argued that they not only deceive but can also be deceived. This seems to be a largely unexplored area of research. Furthermore, the perspective offered here suggests some specific strategies, such as the use of certain models developed in the context of animal deception so as to better understand plants. If assuming that plants make decisions suggests that "theories of decision making and optimal behavior developed for animals and humans can be applied to plants" (Schmid 2016), our results suggest that scientific models currently used to understand animal deception could also be employed in the context of plants. For instance, one might seek to apply models of negative frequency-dependent selection (derived from work on Batesian mimicry) or assess how learning or memory constraints in plants are relevant for the evolution of deception.

Recognizing the existence of deceived plants not only has interesting consequences for our understanding on plants, but it has also a significant impact on our view of deception in general. Suppose that assuming the theory of deception put forward in Section 2.2 and the existence of deceived plants actually delivers important and original results. That would have an impact on our confidence on this general approach to deception. All other things being equal, if a particular theory leads to a progressive research project, that should count as a powerful motivation for embracing this theory.

Finally, I think recognizing the existence of deceiver and deceived plants also probably has more specific consequences in many theoretical debates, such as the nature of self-deception. Exploring these consequences, however, is work for future research.

Acknowledgements

I would like to thank Josep Corbí, Marco Facchin, Gabriele Ferretti, Vladimir Krstic, Sergi Rosell, Peter Schulte, Nicholas Shea, Antonella Tramacere, Chon Tejedor, Sergi Valor, Markus Wild and the audiences at the conference "Green Intelligence? Debating Plant Cognition"' (Basel, October, 2021) and the "Workshop on the Representational Penumbra" (Valencia, May 2023) for their helpful suggestions and criticisms. The research conducive to this chapter was made possible by the projects "The Representational Penumbra" (PID2021-127046NA-I00) funded by the Ministerio de Ciencia, Innovación y Universidades and the projects "Lying and Deception" (CIGE/2021/160) and "Deceptive Representations" (CISEJI/2023/51) funded by the Conselleria d'Educació, Universitats i Ocupació, Generalitat Valenciana.

Notes

1 This chapter focuses on plant *cognition*, rather than on plant *minds*. Although there is of course a close relationship between these concepts, addressing the latter would raise a whole range of issues that cannot be tackled here. I would like to thank Gabrielle Ferreti for showing that this clarification was required.

2 The distinction between direct and indirect strategies is not clear-cut, and it is not meant to capture exclusive approaches. Rather, I think of them as stressing or focusing on different aspects.

3 Interestingly, the word employed to talk about the plant's reaction to VOCs – "priming" – primarily derives from psychology and refers to the phenomenon whereby a stimulus affects the organism's future reaction to a subsequent stimulus, without this effect being conscious or intentional. So, even though communication falls short of providing evidence for intelligence, I take it that many people take the connection to be suggestive.

4 Although I will use the expressions "sender" and "receiver", note that deception need not involve signals (see Artiga and Paternotte 2018, forthcoming; Krstic, forthcoming).

5 Of course, one could interpret the emission of VOCs differently, e.g. as involving some distant form of manipulation along the lines of Krebs and Dawkins (1978) (see Rendall and Owren 2013). However, this "influence" view of communication has few adherents nowadays and, in any case, adopting this perspective would imply taking a very controversial stand on communication in general (Stegmann 2013).

6 An interesting question in the whitefly case is: Who plays the role of deceiver? Is it the whitefly, the plant, both of them or something else? Note that, whichever turns out to be the right answer, this question does not affect my main point, namely, that plants are deceived.

References

Agrawal, Anurag. A. 1999. "Induced Plant Defense: Evolution of Induction and Adaptive Phenotypic Plasticity." In *Inducible Plant Defenses Against Pathogens and Herbivores: Biochemistry, Ecology, and Agriculture*, edited by Anurag A. Agrawal, S. Tuzun, and E. Bent, 251–68. St. Paul, MN (USA): American Phytopathological Society Press.

Agrawal, Anurag A., Jeffrey Conner, Marc Johnson, and Roger Wallsgrove. 2002. "Ecological Genetics of an Induced Plant Defense Against Herbivores: Additive Genetic Variance and Costs of Phenotypic Plasticity." *Evolution; International Journal of Organic Evolution* 56 (11): 2206–13. https://doi.org/10.1111/j.0014-3820.2002.tb00145.x.

Anders Nilsson, L. 1992. "Orchid Pollination Biology." *Trends in Ecology & Evolution* 7 (8): 255–59. https://doi.org/10.1016/0169-5347(92)90170-G.

Artiga, Marc. 2021. "Bacterial Communication." *Biology & Philosophy* 36 (4): 39. https://doi.org/10.1007/s10539-021-09814-1.

Artiga, Marc, and Cédric Paternotte. 2018. "Deception: A Functional Account." *Philosophical Studies* 175 (3): 579–600. https://doi.org/10.1007/s11098-017-0883-8.

Artiga, Marc, and Cédric Paternotte. forthcoming. "Deception as Mimicry." *Philosophy of Science*. https://doi.org/10.1017/psa.2023.156.

Baluška, František, Dieter Volkmann, and Stefano Mancuso, eds. 2006. *Communication in Plants: Neuronal Aspects of Plant Life*. Berin, Heidelberg: Springer-Verlag. https://doi.org/10.1007/978-3-540-28516-8.

Baluška, František, and Stefano Mancuso, eds. 2009. *Signaling in Plants. Signaling and Communication in Plants*. Berlin, Heidelberg: Springer-Verlag. https://doi.org/10.1007/978-3-540-89228-1.

Bassel, George W. 2018. "Information Processing and Distributed Computation in Plant Organs." *Trends in Plant Science* 23 (11): 994–1005. https://doi.org/10.1016/j.tplants.2018.08.006.

Benitez-Vieyra, Santiago, Natalie Hempel de Ibarra, Anna M. Wertlen, and Andrea A. Cocucci. 2007. "How to Look Like a Mallow: Evidence of Floral Mimicry Between Turneraceae and Malvaceae." *Proceedings of the Royal Society B: Biological Sciences* 274 (1623): 2239–48. https://doi.org/10.1098/rspb.2007.0588.

Boughton, Anthony, Kelli Hoover, and Gary Felton. 2005. "Methyl Jasmonate Application Induces Increased Densities of Glandular Trichomes on Tomato, Lycopersicon Esculentum." *Journal of Chemical Ecology* 31 (October): 2211–16. https://doi.org/10.1007/s10886-005-6228-7.

Calvo, Paco, and Fred Keijzer. 2011. "Plants: Adaptive Behavior, Root-Brains, and Minimal Cognition." *Adaptive Behavior* 19 (3): 155–71. https://doi.org/10.1177/1059712311409446.

Calvo, Paco, Monica Gagliano, Gustavo M Souza, and Anthony Trewavas. 2020. "Plants Are Intelligent, Here's How." *Annals of Botany* 125 (1): 11–28. https://doi.org/10.1093/aob/mcz155.

Cooper, W. R., and F. L. Goggin. 2005. "Effects of Jasmonate-Induced Defenses in Tomato on the Potato Aphid, Macrosiphum Euphorbiae." *Entomologia Experimentalis Et Applicata* 115 (1): 107–15. https://doi.org/10.1111/j.1570-7458.2005.00289.x.

Deroy, Ophelia. 2019. "Categorising Without Concepts." *Review of Philosophy and Psychology.* 10 (3): 465–78. https://doi.org/10.1007/s13164-019-00431-2.

Dolgin, Elie. 2019. "The Secret Social Lives of Viruses." *Nature* 570 (7761): 290–2. https://doi.org/10.1038/d41586-019-01880-6.

Farmer, Edward. 2001. "Surface-to-air Signals." *Nature* 411 (6839): 854–6. https://doi.org/10.1038/35081189.

Firn, Richard. 2004. "Plant Intelligence: An Alternative Point of View." *Annals of Botany* 93 (4): 345–51. https://doi.org/10.1093/aob/mch058.

Font, Enrique. 2019. "Mimicry, Camouflage and Perceptual Exploitation: The Evolution of Deception in Nature." *Biosemiotics* 12 (1): 7–24. https://doi.org/10.1007/s12304-018-9339-6.

Gagliano, Monica. 2017. "The Mind of Plants: Thinking the Unthinkable." *Communicative & Integrative Biology* 10 (2): e1288333. https://doi.org/10.1080/19420889.2017.1288333.

Glazebrook, Jane. 2005. "Contrasting mechanisms of defense against biotrophic and necrotrophic pathogens." *Annual Review of Phytopathology* 43: 205–27. https://doi.org/10.1146/annurev.phyto.43.040204.135923.

Godfrey-Smith, Peter. 2020. *Metazoa: Animal Life and the Birth of the Mind.* London: HarperCollins.

Gross, Michael. 2016. "Could Plants Have Cognitive Abilities?" *Current Biology* 26 (5): R181–84. https://doi.org/10.1016/j.cub.2016.02.044.

Heil, Martin, and Ian T. Baldwin. 2002. "Fitness Costs of Induced Resistance: Emerging Experimental Support for a Slippery Concept." *Trends in Plant Science* 7 (2): 61–67. https://doi.org/10.1016/S1360-1385(01)02186-0.

Kessler, André, Michael B. Mueller, Aino Kalske, and Alexander Chautá. 2023. "Volatile-mediated Plant–plant Communication and Higher-level Ecological Dynamics." *Current Biology* 33 (11): R519–29. https://doi.org/10.1016/j.cub.2023.04.025.

Krebs, J. R., and R. Dawkins. 1978. "Animal Signals Mind-Reading and Manipulation." In *Behavioral Ecology: An Evolutionary Approach*, edited by J. R. Krebs and N. B. Davies, 380–402. Blackwell Scientific Publications. Oxford, UK.

Krstić, Vladimir. forthcoming. "We Should Move on From Signalling-Based Analyses of Biological Deception." *Erkenntnis* 1–21. https://doi.org/10.1007/s10670-023-00719-x

Lee, Jonny. 2023. "What Is Cognitive About "Plant Cognition"?" *Biology and Philosophy* 38 (3): 1–21. https://doi.org/10.1007/s10539-023-09907-z.

Mandelbaum, Eric. 2017. "Seeing and Conceptualizing: Modularity and the Shallow Contents of Perception." *Philosophy and Phenomenological Research.* 97 (2): 267–83. https://doi.org/10.1111/phpr.12368.

Maynard-Smith, John, and David Harper. 2004. *Animal Signals.* Oxford Series in Ecology and Evolution. Oxford and New York: Oxford University Press.

Miller, M. B., and B. L. Bassler. 2001. "Quorum Sensing in Bacteria." *Annual Review of Microbiology* 55: 165–99. https://doi.org/10.1146/annurev.micro.55.1.165.

Milikan, Ruth. 1984. *Language, Thought, and Other Biological Categories.* Cambridge, MA: MIT Press.

Millikan, Ruth. 2017. *Beyond Concepts.* New York: Oxford University Press.

Neander, Karen. 2017. *A Mark of the Mental: A Defence of Informational Teleosemantics.* Cambridge, MA: MIT Press.

Oelschlägel, Birgit, Matthias Nuss, Michael von Tschirnhaus, Claudia Pätzold, Christoph Neinhuis, Stefan Dötterl, and Stefan Wanke. 2015. "The Betrayed Thief – the Extraordinary Strategy of Aristolochia Rotunda to Deceive Its Pollinators." *New Phytologist* 206 (1): 342–51. https://doi.org/10.1111/nph.13210.

Papenfort, Kai, and Bonnie L. Bassler. 2016. "Quorum Sensing Signal-Response Systems in Gram-Negative Bacteria." *Nature Reviews. Microbiology* 14 (9): 576–88. https://doi.org/10.1038/nrmicro.2016.89.

Peter, Craig I., and Steven D. Johnson. 2008. "Mimics and Magnets: The Importance of Color and Ecological Facilitation in Floral Deception." *Ecology* 89 (6): 1583–95. https://doi.org/10.1890/07-1098.1.

Rendall, Drew, and Michael J. Owren. 2013. "Communication Without Meaning or Information: Abandoning Language-Based and Informational Constructs in Animal Communication Theory." In *Animal Communication Theory: Information and Influence*, edited by Ulrich E. Stegmann, 151–88. Cambridge: Cambridge University Press. https://doi.org/10.1017/CBO9781139003551.010.

Roy, Bitty A., and Alex Widmer. 1999. "Floral Mimicry: A Fascinating yet Poorly Understood Phenomenon." *Trends in Plant Science* 4 (8): 325–30. https://doi.org/10.1016/s1360-1385(99)01445-4.

Schaefer, H., and Graeme Ruxton. 2009. "Deception in Plants: Mimicry or Perceptual Exploitation?" *Trends in Ecology & Evolution* 24 (September): 676–85. https://doi.org/10.1016/j.tree.2009.06.006.

Schiestl, Florian P. 2005. "On the Success of a Swindle: Pollination by Deception in Orchids." *Die Naturwissenschaften* 92 (6): 255–64. https://doi.org/10.1007/s00114-005-0636-y.

Schmid, Bernhard. 2016. "Decision-Making: Are Plants More Rational Than Animals?" *Current Biology* 26 (14): R675–78. https://doi.org/10.1016/j.cub.2016.05.073.

Scott-Phillips, Thomas C. 2014. *Speaking Our Minds: Why Human Communication is Different, and How Language Evolved to Make it Special*. London: Red Global Press

Shea, Nicholas. 2018. *Representation in Cognitive Science*. New York: Oxford University Press.

Simmons, Aaron, and Geoff Gurr. 2004. "Trichome-Based Host Plant Resistance of Lycopersicon Species and the Biocontrol Agent Mallada Signata: Are They Compatible?" *Entomologia Experimentalis Et Applicata* 113 (November): 95–101. https://doi.org/10.1111/j.0013-8703.2004.00210.x.

Stegmann, Ulrich, ed. 2013. *Animal Communication Theory: Information and Influence*. Cambridge: Cambridge University Press.

Stout, Michael J., and Sean S. Duffey. 1996. "Characterization of Induced Resistance in Tomato Plants." *Entomologia Experimentalis Et Applicata* 79 (3): 273–83. https://doi.org/10.1111/j.1570-7458.1996.tb00835.x.

Thaler, Jennifer S., Parris T. Humphrey, and Noah K. Whiteman. 2012. "Evolution of Jasmonate and Salicylate Signal Crosstalk." *Trends in Plant Science, Special Issue: Specificity of plant–enemy interactions*, 17 (5): 260–70. https://doi.org/10.1016/j.tplants.2012.02.010.

Tomasello, Michael. 2008. *Origins of Human Communication*. Cambridge, MA: MIT Press.

Traw, M. Brian, and Joy Bergelson. 2003. "Interactive Effects of Jasmonic Acid, Salicylic Acid, and Gibberellin on Induction of Trichomes in Arabidopsis." *Plant Physiology* 133 (3): 1367–75. https://doi.org/10.1104/pp.103.027086.

Trewavas, Anthony. 2003. "Aspects of Plant Intelligence." *Annals of Botany* 92 (1): 1–20. https://doi.org/10.1093/aob/mcg101.

Trewavas, Anthony. 2009. "What Is Plant Behaviour?*." *Plant, Cell & Environment* 32 (6): 606–16. https://doi.org/10.1111/j.1365-3040.2009.01929.x.

Walling, Linda L. 2008. "Avoiding Effective Defenses: Strategies Employed by Phloem-Feeding Insects1." *Plant Physiology* 146 (3): 859–66. https://www.proquest.com/docview/218612164/citation/48A380679BAB4BFEPQ/1.

Witzany, Guenther, and František Baluška, eds. 2012. *Biocommunication of Plants. Signaling and Communication in Plants*. Berlin, Heidelberg: Springer-Verlag. https://doi.org/10.1007/978-3-642-23524-5.

Zhang, Peng-Jun, Jia-Ning Wei, Chan Zhao, Ya-Fen Zhang, Chuan-You Li, Shu-Sheng Liu, Marcel Dicke, Xiao-Ping Yu, and Ted C. J. Turlings. 2019. "Airborne Host–Plant Manipulation by Whiteflies via an Inducible Blend of Plant Volatiles." *Proceedings of the National Academy of Sciences* 116 (15): 7387–96. https://doi.org/10.1073/pnas.1818599116.

3 Are Plants Representational Systems?

Peter Schulte

3.1 Introduction

In 1926, an unusual film had its premiere at the Piccadilly movie theater in Berlin. It was called *Das Blumenwunder* ("The flower miracle") and contained long time-lapse sequences of growing tobacco plants, pole-climbing passionflowers, unfolding fern leaves and flowering orchids. Audiences had never seen anything like it, and the reactions were enthusiastic. "The natural impression that the plant is soulless disappears completely", wrote the German philosopher Max Scheler after seeing it, describing the tendrils of a passionflower as "[reaching] desperately into the void" and as showing "satisfaction after finding a pole". Theodor Lessing, another philosopher, went even further, crediting the passionflower with a primitive form of "self-consciousness".[1] Many people who watch documentaries like *The Private Life of Plants* (1995),[2] *Kingdom of Plants* (2012) or *The Green Planet* (2022) have similar reactions today.

What we should make of these reactions, however, is unclear. Does time-lapse photography help us to overcome our habitual "plant blindness" and recognize plant behavior for what it really is?[3] Or does it trick us into attributing mind-like properties to completely mindless organisms because of superficial behavioral similarities?

Of course, when we think about these issues today, we have much more to go on than just time-lapse photography. In recent decades, many remarkable types of plant behavior have been discovered and intensely studied, including root tropisms (Muthert et al. 2020), shade avoidance behavior (Pierik and Testerink 2014), responses to herbivores (Wu and Baldwin 2010), responses regulated by circadian rhythms (Gardner et al. 2006) and plant communication (Ninkovic, Markovic, and Rensing 2021). Still, the fundamental question remains whether or not (some of) the processes that generate these behaviors can be legitimately characterized in psychological terms – as "cognitive", "perceptual" or "representational", as involving some form of "reasoning" or "decision-making", or perhaps even as "conscious".

DOI: 10.4324/9781003393375-5

In this chapter, I will not try to address this question in its fully general form. Instead, I will focus on one key aspect of it: the *plant representation question* (as I call it). This question can be put as follows: can (some) plants be legitimately characterized as representational systems? Or, equivalently: do (some) plants possess representational states?

Four points of clarification are in order here. First, as the question already suggests, I will presuppose a representationalist framework in this chapter. In other words, I will assume that human beings, as well as many other animals, do in fact possess representational states; the question I will investigate is whether plants do so, too. This representationalist paradigm is not uncontroversial, but it continues to be the mainstream view in contemporary philosophy and cognitive science, so I contend that it is worthwhile to explore the topic of plant behavior from this vantage point.[4]

Second, I will adopt a definition of the term "representational state" (or "representation", for short) that is fairly standard. To a first approximation, a representation is to be understood as an internal state that represents how the world is or how it is supposed to be. In more technical terms, a representation is an internal state that has semantic content and that can be correct or incorrect, true or false, fulfilled or unfulfilled (partly) in virtue of having this content. Paradigm examples of representations in this sense are beliefs, desires and other propositional attitudes, but also (I would argue) perceptual states.[5] Accordingly, a subject's belief that wheat is a flowering plant is viewed as an internal state with the content <wheat is a flowering plant> (a state that is true iff wheat is, in fact, a flowering plant); likewise, a subject's visual perception of a red sphere is considered to be an internal state with (something like) the content <that's a red sphere> (a state that is correct iff the perceived object is, in fact, a red sphere).[6]

Third, a comment on the connection between representation and cognition. Some theorists define cognition in terms of representation, others maintain on *a posteriori* grounds that all cognitive processes are representational. On these views, the following arguments carry over to the question of whether there are *cognitive* states and processes in plants. However, both ways of connecting cognition with representation are controversial and in need of further discussion, so I will not take a stand on the question of plant cognition in this chapter.

Fourth, a remark on the relationship between representation and consciousness: I contend that the ascription of representational states to a system does *not* entail the ascription of consciousness of any kind, whether it be phenomenal consciousness, access consciousness or reflexive consciousness.[7] The following discussion is thus entirely independent of questions about the possibility of plant consciousness.

With these clarifications in place, here is, in a nutshell, the answer to the plant representation question that I will argue for in the following: we do not yet know whether plants are representational systems, but we have reason to be skeptical of some of the more sanguine assessments of their representational capacities that can be found in the literature.

The plan of this chapter is as follows. In Section 3.2, I develop a theoretical framework for investigating the plant representation question which forms the basis for the subsequent discussion. In Section 3.3, I go on to criticize some widespread "easy arguments" for the claim that plants are representational systems, and in Section 3.4, I critically examine some more sophisticated arguments for this claim. The results of my discussion are summarized in Section 3.5.

3.2 A Framework for Investigating the Plant Representation Question

The plant representation question, as I have formulated it in the last section, raises a familiar methodological problem: the debate about this question is only interesting if it is more than a mere verbal dispute. From the examples cited in the last section, it is already clear that there must be internal states in plants that are influenced by environmental conditions and that cause adaptive responses like, e.g., gravitropic responses in roots or shade-avoidance responses in shoots. Whether or not we decide to apply the *word* "representation" to these internal states, or (equivalently) whether we *say* that they have the "semantic content" that p when they are normally caused by some environmental state p, is not an important issue. Which terms we use to talk about these states is, in the end, just a matter of linguistic convention.[8] So, how can we prevent the debate about the representation question from degenerating into a futile dispute over words?

A promising first step toward solving this methodological problem is the adoption of an explanatory conception of representation. According to this conception, which is (either implicitly or explicitly) presupposed by many participants in the debate, we should ascribe representational states to an organism, or any other system, if and only if such ascriptions have some distinctive "explanatory benefit" or "explanatory purchase" (see, e.g., Sterelny 1995; Ramsey 2007; Burge 2010; Rescorla 2013; Schulte 2015; Gladziejewski and Milkowski 2017; Shea 2018; Morgan 2018). Or, as I like to put it: we should ascribe representational states to a system S iff these ascriptions can play a role in genuine representational explanations of some of S's behavioral capacities.[9]

When I speak of "representational explanations" here, I do not mean any explanation that is couched in representational terms (otherwise, the methodological problem would immediately re-emerge); I mean a specific *kind* of explanation that is widely used in psychology, ethology and cognitive neuroscience. Explanations of this kind generally aim to account for some capacity of an organism (e.g., the rat's capacity to navigate a maze, the toad's capacity to distinguish prey from non-prey, or the macaque's capacity to grasp an object), and they do so, very roughly, by (i) specifying how numerous families of representational states systematically interact with each other, and by (ii) characterizing these interactions as inferences or mathematical operations of some kind. Much more could be said about these explanations,[10] but a rough grasp of their nature is sufficient in order to appreciate the force of the following arguments.

To illustrate what it means to adopt an explanatory notion of representation, let us consider the contrast between two very different systems. The first system is the firing mechanism of an ordinary rifle (see Ramsey 2007, 136–9; Shea 2018, 30). This mechanism works as follows: when the trigger of the rifle is pulled, the firing pin moves and strikes the primer, which causes a shot to go off. In other words, the pulling of the trigger initiates a linear causal process that leads to the firing of a shot. The mechanism is thus quite simple. Nevertheless, it is possible, at least in principle, to characterize it in representational terms: we could say that, whenever the trigger is pulled, the rifle subsequently *represents* (or *perceives*) that the trigger has been pulled and responds by firing a shot. In accordance with this description, the movement of the firing pin could be characterized as the "vehicle" of the representational state in question.

On the face of it, however, it seems clear that this representational characterization *cannot* play a role in a genuine representational explanation of the rifle's capacity to fire (or, to be precise, its capacity to fire when triggered). The representational model that treats the firing pin's movements as representations of trigger pulls would never be taken seriously by firearm experts, and it is certainly nowhere to be found in the literature on firearm engineering. This suggests that the representational model is "explanatorily idle", a mere *façon de parler* (Ramsey 2007, 139; Shea 2018, 30). Hence, if we adopt the explanatory conception of representation, we have strong reasons to reject representational characterizations of ordinary firearms.

Things look very different when we turn to our second example: the visual system of human beings. Whether we look at David Marr's (1982/2010) classical computational model of vision, the two-streams hypothesis by David Milner and Melvyn Goodale (2006), the comprehensive account of vision developed by Stephen Grossberg (2021) or any other mainstream theory of human vision, we always find that the visual system is described by a rich representational model that seems to have considerable explanatory value. Hence, given the explanatory notion of representation (and the plausible assumption that these mainstream theories are not completely off track), we are justified in attributing representational states to the human visual system.

The explanatory conception of representation thus seems to support ascriptions of representational states to the human visual system and while being incompatible with ascriptions of representational states to the rifle.

This is something that most theorists can agree on. The crucial question, however, especially in the present context, is this: *what makes it the case that some systems (like the visual system of humans) are proper targets for genuine representational explanations, and thus qualify as representational, while other systems (like the rifle's trigger mechanism) do not?* An answer to this question would specify the features which separate representational from non-representational systems and thus enable us to decide, given sufficient empirical information, whether there are plants that belong to the first category.

Unfortunately (but unsurprisingly), there is no consensus about how to answer this question. Very different accounts of the distinction between representational and non-representational systems (or accounts of "representational status", as I will also say) have been proposed in the literature and are still under discussion (Lloyd 1989; Sterelny 1995; Burge 2010; Schulte 2015; Artiga 2016; Gladziejewski and Milkowski 2017; Adams 2018; Shea 2018; Orlandi 2020). If we aim to make progress on the plant representation question, it is plain that we cannot stay neutral between them. Hence, I will lay out what I take to be the most promising theory of representational status and use this theory later on (in Sections 3.3 and 3.4) as the basis for evaluating arguments for representational states in plants. However, I will also occasionally discuss how these arguments fare if other theories of representational status are adopted.

So, without further ado, here is the account of representational status that I find most convincing (and which happens to be my own).[11] To a first approximation, this account runs as follows: when we consider systems with internal processes that have the function of mediating between external stimuli (input) and overt responses (output), the crucial difference between those systems that are representational and those that are not is *a difference in the complexity of the information-processing.*[12] The processing in the human visual system is very complex indeed, involving mechanisms that integrate many types of information from a variety of different sources, and transform this information in many different ways. According to my view, this is why it is possible to construct models of the visual system that are genuinely representational and have real explanatory value.[13] By contrast, the "information processing" in the rifle (if we want to call it that) is simply a case of linear information transfer. All that happens here is that later elements in the causal chain carry information about earlier elements. Consequently, representational descriptions of the process seem utterly pointless. Given the explanatory conception of representation, this entails that the human visual system qualifies as representational while the rifle's firing mechanism does not.

This account yields plausible results, is wholly non-mysterious and does not introduce any arbitrary or unmotivated distinctions. Moreover, I would argue that it preserves what is plausible in many rival theories of representational status, since these theories often focus on features that *contribute* to the complexity of the information processing (with different theories focusing on different features).[14] Finally, it is an account that seems to fit very well with the explanatory practice of the empirical sciences, where we can observe that representational models are taken more and more seriously as the information-processing complexity of the modeled systems increases. Thus, I contend that it forms an excellent basis for our investigation of the plant representation question.

There is, however, one important implication of my account that should be discussed before we proceed. It is clear on reflection that information-processing complexity must be a *gradual* matter (regardless of how exactly this notion is spelled out). Hence, if the distinction between representational and non-representational systems is analyzed in terms of information-processing complexity, it stands to reason that this distinction is also best construed as gradual. Consequently, instead of treating some systems as representational *simpliciter* and all others as non-representational, it seems that we should ascribe *degrees of representationality* to different kinds of systems, depending on their degree of information-processing complexity: we should say that some systems are highly representational, some fairly representational, some moderately representational, some barely representational, and so on.[15] (Plausibly, we should also say that some systems are wholly non-representational, i.e., have a zero degree of representationality.[16])

I think that this line of thought is essentially correct: We should indeed adopt this gradual notion of representationality and accept it as the primary notion when it comes to questions of representational status. Hence, the basic plant representation question is not "are plants representational systems?" but "*how representational* are plants?" (or, equivalently, "what is their *degree* of representationality?").

However, it would be a laborious endeavor to rephrase the whole discussion about plants in terms of representationality. Hence, I think it is useful for the purposes of this chapter to define a secondary notion of representation *simpliciter*. As I see it, what participants in the debate usually mean when they say that plants have representations is that they have representations *in roughly the sense in which (typical) insects, fishes and amphibians have representations*. The reason why insects, fishes and amphibians can serve as suitable points of reference here is the fact that these animals are the simplest organisms (in terms of the information-processing they exhibit) where most theorists agree that genuine representational models have some non-negligible explanatory value. Accordingly, I propose the following provisional definition of "representational system" *simpliciter*: a system S is representational *simpliciter* iff S has (at least) a degree of representationality that is comparable to the degree typically found in insects, fishes and amphibians (a "moderately high degree", as we might say).[17] The "plant representation thesis", which says that plants are representational systems, will from now on be understood in accordance with this definition.

To investigate this thesis, we need to take a closer look at the complexity of the information processing in plants and ask whether it is comparable to the complexity of the information-processing that we find in insects, fishes or amphibians. This is, in a nutshell, what I aim to do in the following. More specifically, I will look at some of the evidence that theorists have cited in order to justify the attribution of representational states to plants and evaluate whether this evidence supports the view that plants possess some fairly complex information-processing mechanisms and, consequently, the plant representation thesis.

Hence, as I have already stated, I will mostly presuppose my account of representational status when evaluating arguments for the plant representation thesis, although I will sometimes also briefly examine whether these arguments look different from the perspective of alternative accounts.

3.3 Why Easy Arguments for the Plant Representation Thesis Fail

Numerous arguments for the plant representation thesis can be found in the literature, although they are not always formulated explicitly as arguments for the claim that plants have representational states. Often, they take the following form: some experiments or observations are presented to support a description of a plant that is either plainly representational or is naturally interpreted as such – e.g., a description which characterizes the plant as *perceiving* that something is the case, as performing acts of *reasoning* or *inference*, or as *deciding* between alternatives. In this section, I will examine a subclass of arguments of this type which I call "easy arguments" and show that they do not provide adequate support for the plant representation thesis.

Easy arguments are found in many popular writings on "plant intelligence" (see, e.g., Chamovitz 2012; Mancuso and Viola 2015; and especially Wohlleben 2016), but they are also sometimes in the background in academic discussions of the topic, as we will see in Section 4. In its basic form, an "easy argument" for the plant representation thesis starts from the observation that a certain plant responds to some variation in its environment by varying its behavior in an adaptive way. To put it schematically, it is observed (i) that a plant normally φ-s when C is the case, but does not φ when C is not the case, and (ii) that φ-ing iff C is the case is beneficial for the plant from an evolutionary point of view. From this, it is concluded that the plant *represents* what is going on in its environment (i.e., whether or not C is the case). The rule of inference employed in this argument may be called a "representation-from-functional-response inference".

Here is a prototypical example of such an argument. It has long been known that orchids of the genus *Catasetum* typically respond to visiting bees by attaching a "pollinium" (a structure containing the plant's pollen) to their backs (Darwin 1862; Nicholson et al. 2008). Obviously, this is a highly adaptive response to the presence of bees, since they serve as pollinators for these orchids. Hence, one might be tempted to conclude from this (via the representation-from-functional-response inference) that *Catasetum* orchids "represent", "perceive" or "recognize" when a bee is present.

The mechanism underlying the plant's adaptive responses, however, is quite simple (see Nicholson et al. 2008). When a bee in search of nectar touches the orchid's "antennae", the impulse is conveyed to a membrane, and if the impulse is strong enough, the membrane is ruptured and the pollinium is released. Hence, what we have here is much like the rifle's firing mechanism described in Section 3.2: a simple, linear causal process leading from the stimulus (touch) to the response (pollinium release). No real information *processing* is involved in bringing about the response, only information transfer.

Consequently, these observations do *not* support the claim that the orchid possesses a moderately high degree of representationality or, equivalently, the claim that the orchid has representational states *simpliciter*. The easy argument for the representational status of *Catasetum* orchids thus fails. This result can be generalized: easy arguments which aim to establish some version of the plant representation thesis *merely* on the basis of adaptive responses to environmental variation cannot succeed, since it is always possible that these adaptive responses are brought about by mechanisms that do not involve any (complex) information-processing. Hence, the representation-from-functional-response inference must be classified as a fallacy.

As already noted, this evaluation is based on the theory of representational status developed in the last section. One might well ask, therefore, how easy arguments look from the standpoint of alternative theories. The short answer is: most other theories yield the same result as my account (e.g., the theories proposed by Lloyd 1989; Sterelny 1995; Burge 2010; Schulte 2015; Gladziejewski and Milkowski 2017; Adams 2018; Shea 2018; Orlandi 2020). However, there is one approach that supports a different verdict, namely *radical liberalism about representation*. Proponents of this approach claim that even internal states that are mere causal intermediates between specific inputs and specific (functional) outputs should be recognized as representations (Artiga 2016, 2022; see also Millikan 1984, ch. 6; Millikan 2009, 405–6). Since easy arguments do indeed establish that there are such states, it is possible to defend these arguments by adopting a radically liberal account of representational status.

However, there are two reasons why this is only cold comfort for defenders of the plant representation thesis. First, it is very doubtful whether radical liberalism is compatible with an explanatory conception of representation. If it is not, then it can be argued that proponents of radical liberalism are simply changing the subject: they are using the term "representation" in a deviant sense, so their theories do not even count as theories of representation (properly speaking). Second, if defenders of the plant representation thesis want to pursue this strategy, they should make their assumptions explicit. They should acknowledge that the account of representational status they rely on to justify the ascription of representational states to plants is an account that also entails, e.g., that the movement of a rifle's firing pin can be described as representing that the trigger has been pulled,[18] or that the internal states of a simple thermostat count as representations of room temperature.[19] Otherwise, their defense of representations in plants is highly misleading.

Let us return to the easy arguments themselves. Not all of them come in the simple form sketched above. Some authors start from the fact that a plant exhibits a functional response to some environmental condition C, as in the basic cases, but instead of merely describing the plant as representing that C is the case, they attribute even more sophisticated representational capacities to the plant.

Here is an example of an easy argument of this kind. In their book *Brilliant Green* (2015), Stefano Mancuso and Alessandra Viola describe the "escape from shade" phenomenon, also known as the "shade avoidance response". This response occurs when a plant finds itself (wholly or partly) in the shade of other plants and it involves, among other things, an increase in stem growth (Pierik and Testerink 2014, 6). Under favorable conditions, this enables the plant to escape from the shade of its competitors. Mancuso and Viola suggest that these observations allow us to draw some striking conclusions:

> The "escape from shade" phenomenon can be seen so well with the naked eye that it was already perfectly familiar in ancient Greece. Yet, [...] its essential significance continues to be ignored or underestimated. What is it that we are actually talking about, after all? Nothing less than a genuine expression of intelligence, which means calculating risk and estimating benefits [...]. The plant's behavior shows that it can plan and use resources to bring about future results [...].
>
> (Mancuso and Viola 2015, 48–9)

Apparently, what Mancuso and Viola want to conclude from the "escape from shade" phenomenon is not merely that plants can represent the presence of shade (or the presence of competitors) but that they can "[cal-culate] risk and benefits" and "plan" for the future!

Yet, from an information-processing perspective, the mechanism responsible for the shade avoidance response may not be much more complex than the pollinium-ejection mechanism of *Catasetum* orchids. In the case of shade avoidance, the plant is reacting to a specific stimulus, namely shade (or more precisely, the shade of other plants, which is characterized by a low ratio of red light to infra-red light; see Pierik and Testerink 2014, 7),[20] and this stimulus causes a biochemical cascade that leads to an increased growth of the plant's stem. For all we know, this may essentially be a case of a (long) linear causal chain connecting a specific stimulus to a specific functional response.[21]

Hence, the mere fact that many plants react to the shade of other plants by growing faster, the fact that "was perfectly familiar in ancient Greece", does nothing to establish that complex information-processing is going on in those plants. It thus fails to justify the ascription of representational states to plants, let alone the ascription of processes of risk calculation and planning. What appears to be going on in the passage by Mancuso and Viola is an illicit inference from the fact that the plant exhibits behavior that maximizes fitness (or, perhaps, its own "welfare") to the conclusion that the plant performs some kind of fitness-maximizing (or welfare-max-imizing) calculation. This ignores the possibility of what Daniel Dennett calls "competence without comprehension": that an organism can be "the beneficiary of a routine that it follows without any comprehension of its rationale" (Dennett 2009, 10063). Accordingly, the argument presented by Mancuso and Viola may be called a "comprehension-from-competence fallacy".

Two clarificatory remarks are in order here. First, I am not precluding the possibility that the "escape from shade" mechanism might turn out to involve some complex information processing after all. My argument is only that given the evidence that I have described here (and that is alluded to by Mancuso and Viola), we have no positive reason to think so. Secondly, I want to emphasize that the complexity that is at issue here is *information-processing complexity*, not mere physical complexity. There is ample evidence that the physical process underlying the "escape from shade" phenomenon is very intricate: the biochemical cascade that connects the stimulus to the response involves many different kinds of photoreceptor proteins, transcription factors, hormones and cell wall-modifying proteins that interact with each other in complex ways (Pierik and Testerink 2014, 7–11; Pierik and de Wit 2014). However, that is *not* the kind of complexity that warrants representational characterizations.

Many other instances of easy arguments can be found in literature, sometimes explicitly stated, and sometimes merely hinted at, but these arguments are no more convincing than the arguments discussed here, for exactly parallel reasons. To put it simply, a system can generate an adaptive response to a specific stimulus (or a range of adaptive responses to a range of specific stimuli) without engaging in any complex information-processing. Hence, easy arguments fail across the board.

3.4 Better Arguments for the Plant Representation Thesis?

How would better arguments for the plant representation thesis look like? In general, we can say that these arguments would draw on behavioral data that are, at least prima facie, best explained by postulating complex inner processes of information integration and transformation. Ideally, these arguments would also provide evidence for the existence of physiological structures that could implement these processes. Arguments which combine behavioral and physiological evidence in this way would indeed lend substantial support to the claim that (some) plants are genuine representational systems.

The arguments for representations in plants that best fit this abstract characterization, at least at first glance, are based on the investigation of root growth in plants. Supporters of the "plant cognition" paradigm (or "plant cognitivists", as I will call them) regularly assert that plant roots exhibit coordinated and flexible responses to a multiplicity of stimuli, which suggests that roots are able to integrate information about these stimuli (Calvo and Keijzer 2011; Trewavas 2014, 122–36; Novoplansky 2019; Lee, Segundo-Ortin, and Calvo 2023), and some plant cognitivists also aim to provide physiological evidence which indicates that the "hardware" required for such complex information processing is present in plant roots (Baluška et al. 2009; Baluška et al. 2010). On this basis, many plant cognitivists defend characterizations of plants that are explicitly representational: Anthony Trewavas claims that "roots construct some kind of *image* of their root environment" (Trewavas 2014, 122; my emphasis), Paco Calvo and Karl Friston maintain that plant roots "constantly assess the (future) acquisition of minerals and water" (Calvo and Friston 2017, 4),[22] and Mancuso and Viola suggest that the root tip "guides the root on the basis of a real calculus that takes into account the plant organism's different local and global needs", describing this region as a "true 'data processing center'" (Mancuso and Viola 2015, 140).

How convincing are these arguments? Let us first look at the physiological evidence. Many plant cognitivists highlight the fact that, in addition to hormonal signaling, there is also electrical signaling in plants (Brenner et al. 2006; Baluška et al. 2009; Calvo 2016, 1326–9). Indeed, action potentials (APs) that resemble those in animal nervous systems are found in many plant species, along with so-called "variation potentials" (VPs) and "system potentials" (SPs), which are specific to plants (for a recent review, see

Choi et al. 2016, 293–4). Apparently, electrical signals of these kinds travel along pathways that stretch the plant's whole body, including the roots, and there is strong evidence that they are involved in the detection of mechanical wounding and other stress-inducing stimuli (Choi et al. 2016, 292).

However, to show that electrical signals are present in plants is one thing, to show that they are involved in complex forms of information processing is quite another. Take, e.g., the Venus flytrap, where the role of electrical signals in the production of the plant's characteristic "snapping behavior" has been extensively studied (Hodick and Sievers 1989; Forterre et al. 2005). As it turns out, this role is a rather simple one: whenever something touches one of the "trigger hairs" located on the trap, an AP is generated, and if two APs are generated within a time span of 20 seconds, the trap snaps shut. From an information-processing perspective, this mechanism has essentially the same (low) level of complexity as the *Catasetum*'s pollinium-ejection mechanism or the firing mechanism of the rifle.[23] And for all we know, the same might be true for the electrical signals involved in the detection of stress-inducing stimuli that are widespread in the plant kingdom: they might serve as system-wide "alarm calls" that trigger appropriate responses in different parts of the plant without giving rise to any complex information-processing.[24] Indeed, some scientists, like the eminent plant biologist Lincoln Taiz and his co-authors (2020), are very skeptical of the idea that electrical signaling in plants has anything to do with sophisticated forms of information integration: "To our knowledge there are no examples in plants of groups of vascular cells (the putative plant nervous system) using electrical signals to process integrative information" (Taiz et al. 2020, 218).

Plant cognitivists also appeal to further physiological evidence to support their claim that complex information-processing is going on in plant roots. They point to the fact that molecules that serve as neurotransmitters in animals have also been found in plants (e.g., serotonin, dopamine and GABA), and contend that plants in general and plant roots in particular contain structures that resemble certain elements of neuronal systems (Brenner et al. 2006; Calvo 2016; Calvo and Trewavas 2020). Most strikingly, František Baluška, Stefano Mancuso and their co-authors have claimed that the so-called "transition zone" in the root tip is a "'brain-like' command centre" containing "plant synapses" (Baluška et al. 2009, 2010), describing their proposal explicitly as an attempt to revive the "'root-brain' hypothesis" floated by Charles and Francis Darwin in 1880 (Baluška et al. 2009).

However, the actual evidence provided for these claims is rather weak. Talk of "plant synapses" in Baluška et al. (2009, 2010) is mainly based on observations of two types: (i) observations which support the hypothesis that auxin (a phytohormone crucial for several types of tropistic behavior) is transported by vesicles within cells and released when these vesicles fuse with the outer cell membrane, much like neurotransmitters are transported and released in animal brains (Baluška et al. 2009, 1123; Baluška et al. 2010, 405–6), and (ii) observations that suggest a possible role for auxin in electrical signaling (Baluška et al. 2010, 406).

This evidence seems insufficient, however, for several reasons. First, the claim that there is vesicle-mediated transport of auxin in plant cells is disputed (Alpi et al. 2007, 136). Secondly, there are, in any case, significant disanalogies between auxin transport in plants and neurotransmitter transport in brains: e.g., auxin is transported from cell to cell over long distances, while neurotransmitters are not (Alpi et al. 2007, 136). Thirdly, and most importantly, even if Baluška, Mancuso and their co-authors are right about auxin transport and the role of auxin in electrical signaling,[25] this does not go very far in establishing their thesis about brain-like command centers in plant roots. When we look at synapses in animal brains, it is clear that there are two facts about them that are essential to their role in complex information-processing: (a) the fact that neurons generally have *specific* synaptic connections to *many other neurons*, receiving inputs from some of them and transmitting outputs to others, and (b) the fact that synapses have different *weights*, so that the activity of the presynaptic neuron can influence the postsynaptic neuron in a variety of different ways.[26] These are the synaptic properties that enable neural networks to transform the information they receive in complex ways (see, e.g., Churchland 2012; Eliasmith 2013). However, as far as I can see, Baluška et al. (2009, 2010) give us no evidence which suggests that so-called "plant synapses" have these crucial properties, and without them, other physiological similarities seem largely irrelevant.[27]

Finally, as Mallatt et al. (2021) have pointed out, the root tip is prima facie implausible as a location for a brain-like structure, since the cells of which it consists are continually replaced by new cells, often at an astonishingly high rate.[28] This contrast sharply with cell replacement rates in animal brains, which are extremely low (see, e.g., Sender and Milo 2021) and must be so in order to make the long-term preservation of weighted synaptic connections possible. The high replacement rates in the root tip thus appear to be "incompatible with the formation of the stable processing networks" that would be required for complex information-processing (Mallatt et al. 2021, 465).

In sum, it seems that there is no substantive physiological evidence for the existence of "plant brains", i.e., of structures in plants that resemble brains or neural networks of animals in relevant respects. Of course, this does *not* allow us to conclude that there is no complex information-processing in plants. For one thing, it is possible that brain-like structures do exist in plants after all, even though we do not currently have good physiological evidence for their existence. Moreover, there is another, perhaps more plausible possibility: it may be that the kind of complex information-processing that is carried out by neural systems in animals is implemented in a completely different way in plants, e.g., by complex forms of hormonal signaling or by a combination of hormonal and electrical signaling.

Hence, it is of prime importance to take a closer look at the behavioral data that are cited in support of the claim that plant roots engage in large-scale information-processing. Paco Calvo takes it as a well-established fact that plant roots "can sample *and integrate* in real time many different biotic and abiotic parameters, such as humidity, light, gravity, temperature, nutrient patches and microorganisms in the soil, and many more [...], with sensory information being transduced via a number of modalities" (Calvo 2016, 1329; my emphasis). Similar claims can be found in the writings of many plant cognitivists (Calvo and Keijzer 2011, 157; Baluška et al. 2009, 1124; Novoplansky 2019, 126). However, since such claims are open to different interpretations, and also somewhat controversial, we need to examine in more detail the empirical evidence that is provided for them.

Let me begin by noting that there is no doubt that root growth is *influenced* by a number of different environmental variables. For instance, when roots encounter nutrient-rich patches in the soil, they often proliferate within them, i.e., they grow more lateral roots in order to enhance nutrient uptake (Hodge 2004). Furthermore, roots exhibit several types of directional responses known as "tropisms". At least five of these tropisms are well-documented: gravitropism (responses to gravity), phototropism (responses to light), hydrotropism (responses to humidity), halotropism (responses to salinity) and thigmotropism (responses to touch). Other tropisms have been proposed but are still subject to some controversy (for a comprehensive review, see Muthert et al. 2020).

Root growth thus depends on a number of different environmental parameters.[29] However, it should be clear that this fact alone does not yet establish that roots *integrate* information about these parameters in the relevant sense. Information integration in this sense requires the existence of an internal mechanism where states that carry information about several different parameters causally interact with each other in significant ways to determine the plant's behavior, and the mere fact that root growth depends on several different environmental parameters does not guarantee that roots employ mechanisms of this kind. After all, it is possible, at least in principle, that the environmental parameters act on root growth independently from each other, through separate pathways, and that the resulting growth response can be determined by simply "adding up" the effects of these parameters (while holding endogenous factors constant).[30]

Moreover, even if the effects are not independent, this does not neces-sarily mean that processes of information integration are involved. To take a simple example, suppose that increased root growth in areas that have been enriched with nutrients is correlated with decreased root growth in other, non-enriched areas (Hodge 2009, 631). This *can* be the result of information-processing (e.g., of a process in which the nutrient-richness of different areas is compared and then, based on this comparison, signals are sent out to stimulate growth in some areas and inhibit it in others); but at least in principle, it can equally well be the result of the fact that increased growth in certain areas depletes some of the plant's internal resources, leading automatically to decreased growth in other areas.[31]

In light of these complications, it stands to reason that claims about information integration in plants need to be adequately justified: their pro-ponents must show that specific types of plant behavior are best explained by postulating inner processes of information integration and, conse-quently, that simpler explanations of these types of behavior can be suc-cessfully ruled out.[32] As a corollary, we can say that arguments like the following must be rejected as illegitimate:

> [R]oots follow gradients of resources and proliferate where they find them. *Thus*, roots construct some kind of image of their root environment.
>
> (Trewavas 2014, 122; my emphasis)

The mere observation that plants show a tropistic (directional) response to the unequal distribution of nutrients, followed by a proliferation response to high nutrient concentrations, cannot by itself support the conclusion that some complex information integration is going on in plants, let alone the claim that plants construct a map-like representation of their environ-ment. Such an argument looks, again, very much like the comprehension-from-competence fallacy that we discussed in the last section.

So, the question is: do we have good evidence for information integra-tion in plants? As a representative example, let us look at a study by Gioia Massa and Simon Gilroy (2003) that is often cited as evidence for the claim that plants integrate information about gravity and touch (see, e.g., Trewavas 2005; Calvo and Keijzer 2011; Calvo et al. 2020). The study is about root growth in *Arabidopsis thaliana*. Massa and Gilroy make the following striking observation: when the root tip of a downward-growing root of *A. thaliana* touches a barrier, it bends into a "step-like" shape and grows sideways until the end of the barrier is reached. At this point, it starts to grow downwards again. Taken together with a number of further experiments, this observation suggests that there is a systematic interac-tion between the root's gravitropic response and the touch response (the "thigmotropic" response). More precisely, it seems that the activity of the touch-sensitive cells in the root cap partially *inhibits* the gravitropic response (Massa and Gilroy 2003, 439). Interestingly, Massa and Gilroy also formulate a hypothesis about the mechanism that underlies this inhibitory effect. It is generally accepted that the sensory cells which control the gravitropic response contain small starch-filled organelles called "statoliths". These statoliths can move around freely in the cell, and since they are heavier than the surrounding cytoplasm, they normally sediment in that part of the cell which is lowest, thus indicating the direction of gravity. Now, according to Massa and Gilroy, activation of the touch-sensitive cells causes a significant reduction in statolith sedimentation rates, which leads to a temporary suppression of the gravitropic response (Massa and Gilroy 2003, 439).

I agree that this is indeed a form of information integration in the relevant sense. It involves internal states that carry information about external parameters interacting with each other to produce plant behavior. However, while this is quite fascinating, we should note that it is still a rather rudimentary form of information integration – a "two-factor" process where two signals interact in a fairly straightforward way to determine the system's response. In the terminology of Chris Reid et al. (2015), it is a "noncompensatory" strategy, where the signals carrying information about one environmental parameter simply *override* the signals that carry information about the other. Such strategies are "computationally simple" (Reid et al. 2015, 45), as the mechanism sketch provided by Massa and Gilroy also indicates.

Consider, as an analogy, a device consisting of a thermostat that is connected to a smoke detector. Let us suppose that both components work as they typically do, but when the smoke detector is activated, it sends out a signal that switches off the thermostat. This device exhibits the same rudimentary form of information-processing that we find in the gravitropism/thigmotropism case: its behavior, i.e., whether and how it regulates the heating system, is determined by the interaction of internal states that carry information about two different external parameters (room temperature and the presence or absence of smoke). Hence, for all we know so far, the plant root may not possess a higher degree of information-processing complexity than this device.

This is a highly significant result, because what is true of the integration of information about gravity and touch may also be true in other cases. And indeed, when we look at how plant roots integrate information about salinity and gravity (Yokawa et al. 2014) or information about the nutrient-richness of different soil patches (Drew and Saker 1978; Novoplansky 2019), then it does seem to be the case that simple, "noncompensatory" models are sufficient to explain most of the data. This opens up the following theoretical possibility: it may be the case that root growth in plants is controlled by a number of independent tropistic mechanisms which interact with each other in informationally minimal ways (for instance, mechanism M_1 inhibiting M_2 under conditions C, mechanism M_2 inhibiting mechanisms M_3 under conditions C* *and* mechanism M_4 under condition C**, etc.). If these interactions are finely tuned by natural selection, such a collection of mechanisms may produce quite flexible and well-adapted behavior in the absence of any genuinely complex information-processing.

So, the question is: do we have reason to think that this rather minimalist model is, in fact, wholly adequate, or is there evidence that something more is going on in plant roots? As I see it, there is indeed *some* evidence for more complex forms of information-processing, but it is quite scarce. Moreover, in the writings of plant cognitivists, it is often buried in heaps of data that are much less compelling.

Here is one of the studies which suggests that the minimalist model cannot account for all forms of plant behavior. It is a study of root growth in *Abutilon theophrasti* (commonly known as velvetleaf or Chinese jute) by James Cahill et al. (2010), which seems to show that *A. theophrasti* integrates information about nutrients and competitors in a way that goes beyond mere inhibitory interactions (see Figure 3.1).[33] First, plants were grown alone, both under conditions where nutrients were uniformly distributed (condition A in Figure 3.1) and under conditions where nutrient distribution was "patchy" (conditions B and C). These plants showed no significant variation in root growth. Second, plants were grown with a competitor in the vicinity, again under conditions of uniform nutrient distribution (condition D) and under conditions of patchy nutrient distribution (conditions E and F). This time, there was a significant difference in root growth: in condition D, both the breadth and length of roots were greatly reduced (relative to conditions A, B and C), but in conditions E and F, the reduction in root breadth was much less pronounced,[34] especially when the nutritious patch was located between both plants (condition E).

Cahill et al. (2010, 1657) conclude from these observations that "plants nonadditively integrate information about both resource and neighbor-based cues in the environment", i.e., that they show responses which cannot simply be explained as the combined effects of two independent mechanisms (or, we might add, as the effects of two mechanisms that only interact in a "noncompensatory" fashion). Indeed, the data suggest that there is a mechanism for root growth which can, on the basis of (i) information about the presence or absence of competitors and (ii) information about the nutrient distribution, select between at least four alternative strategies. If there are no competitors, the mechanism opts for *normal root growth* (see A, B, C). If there is a competitor and nutrient distribution is uniform, it triggers an *avoidance response* (great reduction

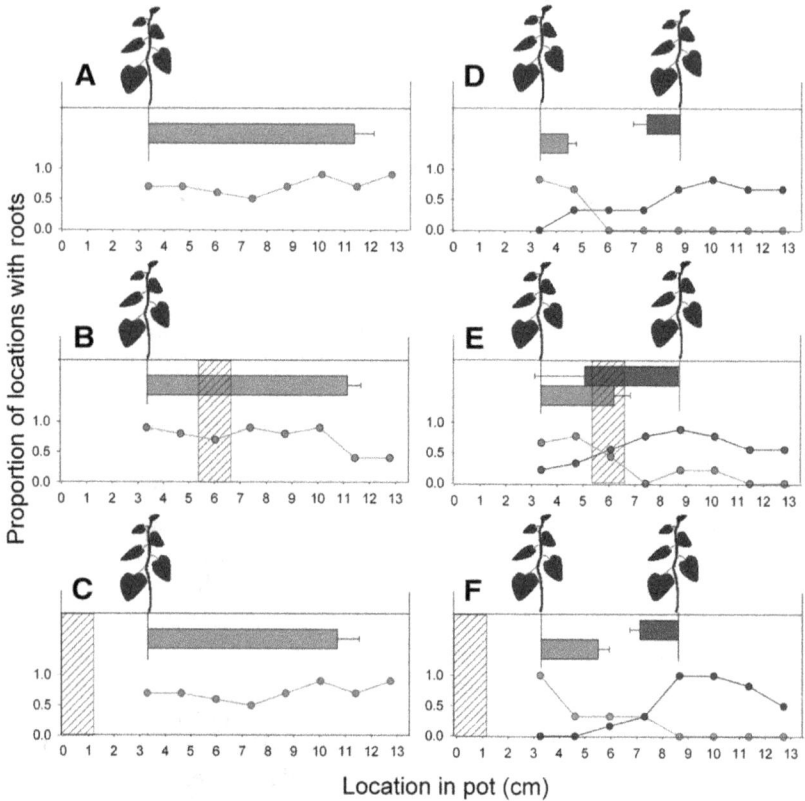

Figure 3.1 Root growth in *Abutilon theophrasti* (from Cahill et al. 2010). The horizontal bars stand for average root breadth, the lines at the bottom indicate how far from the plant the roots extend.

in root growth, see D). If there is a competitor and nutrient distribution is patchy, but the patch does not lie in the same direction as the competitor, a *moderately competitive response* is triggered (less reduction in root growth, see F). Finally, if there is a competitor and nutrient distribution is patchy, and the patch does lie in the same direction as the competitor, a *strongly competitive response* is selected (much less reduction in root growth, see E).

To sum up, there is some evidence for information integration in plants that goes beyond simple "noncompensatory" processes of inhibition. However, this evidence is thin on the ground. Apart from Cahill et al. (2010), there are only a few other studies that appear to provide clear evidence of the existence of more complex internal processes of information integration.[35] Furthermore, the evidence we have is preliminary, and in most cases, the physiological mechanisms underlying these more complex forms of information integration are not fully understood.[36] Finally, and most importantly, the evidence we have so far, even if it is corroborated by further studies, is consistent with information-processing in plants that is still much less complex than the information-processing we find in the nervous systems of insects, fishes and amphibian.[37] Hence, the bold claims of some plant cognitivists notwithstanding, we do not currently have strong reasons to believe that the plant representation thesis is true, and even some reason to doubt that it is.

Of course, this conclusion is ultimately based on the theory of representational status that I have advocated in Section 3.2. So, it is again interesting to ask whether alternative theories would yield the same verdict. In general, we can say that there are a number of more restrictive (non-liberal) theories of representational status that also entail, given the evidence laid out in this chapter, that we have no reason to believe that plants are representational systems. This is true, e.g., for the theories defended by Tyler Burge (2010, 2022), Fred Adams (2018) and Donald Davidson (1982). Obviously, radically liberal theories yield a different verdict, for the reasons laid out in the above discussion of "easy arguments" (Section 3.3). However, what is different now is that there are also some *other* liberal theories which seem to support the plant representation thesis.

Take, for instance, Nick Shea's (2018) theory of representational status. This theory is not a version of radical liberalism: Shea rejects the claim that mere causal intermediates between specific inputs and specific (functional) outputs qualify as representational states. However, he accepts that mechanisms that are only *slightly* more complex do involve representations. For instance, he considers the hypothetical case of a plant with an internal state R that leads to the production of a specific (functional) output, namely flower-closing behavior, but that can be (normally) caused by a drop in temperature *or* by a dimming of the light, i.e., by *two different* inputs which both indicate that evening has arrived (Shea 2018, 213–14). According to Shea, this state R does count as a representation. Shea's example is hypothetical, but there certainly are mechanisms of this level of complexity in plants, e.g., defense mechanisms against herbivores that can be triggered by mechanical wounding or by certain chemical cues (Wu and Baldwin 2010).

It is thus possible to defend the plant representation thesis on the basis of Shea's account of representational status. However, the points that I have made in connection with the radically liberal view also apply in this case. First, it is unclear whether Shea's account is compatible with the explanatory conception of representation (see Rescorla 2021). Secondly, if plant cognitivists want to defend their view in this way, they should be upfront about it. They should acknowledge that they rely on a theory of representational status that does not only entail that plants have representations but also that rather simple mechanical devices do (e.g., a firing mechanism that can be activated by pulling a trigger *or* by sending an electronic signal via remote control). Accordingly, the gist of their argument would not be that plants are much more sophisticated than we thought they were but that organisms need not be very sophisticated in order to qualify as possessing representational states.

3.5   Conclusion

Are plants representational systems? If anything has become clear in the course of this chapter, it is that this question is very difficult to answer. The first problem is that we cannot proceed in a "theory-neutral" manner but must instead rely on some theory of representational status. For this reason, I argued in the first part of the chapter for a gradualist theory of representational status which says that a system's degree of representationality is grounded in the complexity of the information-processing that is going on in that system. Moreover, I proposed to understand a representational system *simpliciter* (for the purposes of this chapter) as a system with a degree of information-processing complexity that is comparable to the degree we find in insects, fishes and amphibians. On this basis, I criticized "easy arguments" for the plant representation thesis as fallacious. I then examined more sophisticated attempts to establish this thesis and argued that these attempts also ultimately fail, although there is *some* evidence that plants may possess a *somewhat* higher degree of information-processing complexity (and, *ipso facto*, a higher degree of representationality) than many of us commonly assume.

Even though we are now in a much better evidential situation than the viewers of "Das Blumenwunder" in 1926, we still face a similar problem. Like the movie images, the scientific data we have leave considerable room for interpretation. It thus remains an open question whether plants are representational systems. However, in my estimate, the evidence we currently possess does not point in that direction.

Acknowledgements

I would like to thank Hannah Altehenger, Adrian Wieczorek, Markus Wild and an anonymous reviewer for helpful comments and suggestions that greatly improved the chapter. I am also grateful to Dominique Hosch and audiences at Basel, Potsdam and Umeå for valuable comments on a presentation on which this chapter is based.

Notes

1 These quotes are from Lindner (2020) (my translations).
2 Note: this excellent BBC documentary should not be confused with *The Secret Life of Plants* (1978), a notorious hodgepodge of fake experiments and New Age drivel.
3 For the purposes of this chapter, I am adopting a liberal notion of behavior that includes everything an organism *does*, in the widest sense of the word. This notion does not differentiate between behavior on the one hand and developmental processes (like growth) on the other. Hence, it is unproblematic to speak of "plant behavior" in this liberal sense.
4 By saying this, I do not mean to deny that it is *also* worthwhile to investigate the topic from other, non-representationalist vantage points (see Calvo 2016; Maher 2017; see also the contributions by Frazier and Lee and Ponkshe in this volume). The debate between representationalists and non-representationalists is still far from being settled, so it is reasonable to pursue these approaches in parallel.
5 It is possible to accept that beliefs and desires are representations while rejecting the claim that perceptual states are. Still, most proponents of the representationalist paradigm agree that perceptual states do qualify as representational. For a helpful survey of the representational view of perception, see Pautz (2021, 93–187).
6 This way of characterizing the content of perceptual states is not meant to suggest that the *format* of perception is language-like rather than iconic. The content of iconic representations can be specified by that-clauses, albeit only in a partial and rough-and-ready manner (hence the qualification "something like").
7 This contention is not uncontroversial, but I consider it to be very well-founded. For discussion of the contrary view that phenomenal consciousness is necessary for representation, see Schulte (2023, 47–50).
8 Analogous problems arise in debates about "plant intelligence" and "plant cognition" (see Cvrčková, Lipavská, and Zárský 2009; Taiz et al. 2019).
9 For an elaboration and defense of this way of spelling out the explanatory conception, see Schulte (ms).
10 See, e.g., Bechtel (2008), Piccinini (2020, ch. 7) and (Shea 2018, ch. 8). For my own take, see Schulte (ms).
11 The following points are more fully developed in Schulte (ms).
12 More precisely, what is crucial is a difference in (what I call) the "structured complexity" of the information-processing, a kind of complexity that excludes giant look-up table architectures as in the "Blockhead" thought experiment (Block 1981). For more on this notion, see Schulte (ms).
13 As these remarks indicate, I see no conflict between explanations in terms of representational states and explanations in terms of information-processing. Quite to the contrary, I hold that complex information-processing explanations have all the hallmarks of representational explanations and should thus be regarded as "covert" representational explanations (i.e., as representational explanations couched in non-representational terms).
14 For instance, some theorists consider perceptual constancies to be necessary for representation (Burge 2010; Schulte 2015), others hold that "stimulus-independent" (memory-like) states are crucial (Orlandi 2020). According to my (present) view, both of these features are *relevant*, since they augment the complexity of the information-processing, but neither of them is *the* decisive feature.

15 Given our definition of representations as states with semantic content (Section 3.1), this means that content properties must also be understood in a gradualist way. We could say, for instance, that states with increasing representationality instantiate properties that are more and more "contentish". Or, to put matters more transparently, we could say that, with increasing information processing complexity, the functional-informational properties of states gradually acquire the character of genuine content properties.

16 On my account, clear cases of wholly non-representational systems are entities that lack any functional traits, and thus (a fortiori) any internal processes that have the function of mediating between external stimuli and overt responses. Some examples are atoms, stones, rivers, clouds and galaxies.

17 More generally, the notion of representation *simpliciter* can be treated as a notion that is (explicitly or implicitly) *context-relative*. Thus a general definition of it might look as follows: system S is representational *simpliciter* (relative to context C) iff S's degree of representationality is sufficiently high (given the standards of context C).

18 To see this, note that the firing mechanism also reacts to a very specific stimulus, a pull of the trigger, by generating an appropriate response, the firing of a shot. The only difference is that the rifle's response is "appropriate" from the point of view of the rifle's designers, while the plant's response is "appropriate" from the point of view of evolution (i.e., adaptive).

19 Note that, on Millikan's view, artifact functions are genuine proper functions, despite the fact that they are not directly determined by processes of natural selection (see, e.g., Millikan 2004, 13). Hence, it is not open to her to deny that the rifle or the thermostat has representational states on the grounds that they do not possess the right kind of function. (The same presumably holds for Artiga, too.)

20 The reason why shadows cast by plants have this characteristic is simply that plant leaves absorb red light while reflecting infra-red light.

21 There is one fact which may indicate that some (minimal) information-processing is going on in this case: the fact that the plant reacts to the *ratio* of red light to far-red light. This suggests that the plant possesses a mechanism for comparing (or "weighing") the amounts of red and far-red light it receives. However, these considerations are obviously orthogonal to the argument by Mancuso and Viola.

22 There are also many other passages in this paper that seem to be explicitly representational, e.g., when Calvo and Friston write that internal states of plants are involved in "inferences" (Calvo and Friston 2017, 2) or that a plant can be described as "passing predictions and prediction errors around its body in the service of minimizing prediction error" (Calvo and Friston 2017, 8). At the same time, however, they seem to distance themselves from representationalism later in the paper (Calvo and Friston 2017, 9), so it is not entirely clear how to interpret their position.

23 There is one relevant difference: the trap-closure mechanism needs *two* APs within 20 seconds to be activated and can thus be described as integrating information about two successive touch stimuli. But obviously, this is only a very simple form of information integration.

24 Also relevant in this context is the following statement from the review article on electrical signaling in plants cited above: "Electrical and Ca^{2+} signals appear to be generated in response to many stimuli, but we currently lack a clear understanding of whether specificity in the stimulus triggering this signaling network is encoded in their spatial and temporal dynamics" (Choi

et al. 2016, 302). In other words, we do not know whether different external stimuli give rise to different types of electrical signals. However, this would be an essential prerequisite for any kind of complex sensory information-processing.

25 In fact, this is a big "if": in a recent paper, Robinson and Draguhn (2021, 3) contend that the process described by Baluška and Mancuso "does not contribute to voltage changes in the plasma membrane", calling its role in electrical signaling into question.

26 A further crucial fact is that these synaptic weights can be *modified* by learning processes.

27 For further important disanalogies, see Robinson and Draguhn (2021).

28 Mallatt et al. (2021, 465) report a replacement rate of 4.7 hours (!) for the transition zone of maize roots.

29 As an experiment discussed later in this section shows, there is also good evidence for the relevance of a seventh parameter: the presence of roots from competing plants.

30 The term "information integration" is sometimes used in a broader sense that does not entail the existence of internal processes of the kind described here. For instance, Moulton, Oliver, and Goriely (2020) seem to employ the term in this broader sense when they say that the "movement [of the plant; PS] *is* an integration of multiple signals" (Moulton, Oliver, and Goriely 2020, 32226; my emphasis) and speak of information integration in cases where they assume that "the effects of multiple stimuli are additive" and the "pathways for signal transduction" are "separate" (Moulton, Oliver, and Goriely 2020, 32232). Hence, we should be careful not to presuppose that any claim about "information integration" in plants that is found in the literature is a claim about information integration in the sense that is relevant here.

31 As a matter of fact, it seems that this simple alternative explanation is *not* correct. It has been convincingly shown that starving rootlets produce signaling molecules that stimulate root growth in other areas (Tabata et al. 2014). However, this does not affect the principled point I am making in this paragraph.

32 Note that appeals to simplicity are unproblematic in this context: dedicated mechanisms for complex information-processing are costly, so if the same results can be achieved by simpler mechanisms, those will be much more likely (given the usual selection pressures).

33 In other words, the study seems to show that the behavior-determining process is "compensatory" in the sense of Reid et al. (2015).

34 There is also a little less reduction in root length, but this effect is very subtle.

35 Examples are studies of responses to nutrient heterogeneity (Cahill and McNickle 2011, 299), of mechanisms for the control of shoot cell elongation (Chaiwanon et al. 2016, 1258) and of the processes that determine flowering time (Kobayashi and Weigel 2007).

36 Cahill and McNickle (2011, 290), for instance, acknowledge that the "processes that determine where plants put their roots within the soil are poorly understood in comparison with analogous aboveground processes in plants [...] and movement patterns among vertebrates".

37 It should be mentioned here that Mancuso and Baluška (2017) have made claims about visual shape-recognition in *Boquila trifoliata* that would, if true, call the point I am making here into question. However, most mainstream scientists are highly skeptical of these claims (and with good reason, I think). For a useful overview of this debate, see Jones (2023).

References

Adams, Fred. 2018. "Cognition Wars." *Studies in History and Philosophy of Science* 68: 20–30. https://doi.org/10.1016/j.shpsa.2017.11.007

Alpi, Amadeo, Nikolaus Amrhein, Adam Bertl, Michael R. Blatt, Eduardo Blumwald, Felice Cervone, Jack Dainty, Maria Ida De Michelis, Emanuel Epstein, Arthur W. Galston et al. 2007. "Plant Neurobiology: No Brain, No Gain?" *Trends in Plant Science* 12 (4): 135–6. https://doi.org/10.1016/j.tplants.2007.03.002

Artiga, Marc. 2016. "Liberal Representationalism: A Deflationist Defense." *dialectica* 70 (3): 407–30. https://doi.org/10.1111/1746-8361.12152

Artiga, Marc. 2022. "Strong Liberal Representationalism." *Phenomenology and the Cognitive Sciences* 21: 645–67. https://doi.org/10.1007/s11097-020-09720-z

Baluška, František, Stefano Mancuso, Dieter Volkmann, and Peter Barlow. 2009. "The 'root-brain' Hypothesis of Charles and Francis Darwin." *Plant Signaling & Behavior* 4 (12): 1121–7. https://doi.org/10.4161/psb.4.12.10574

Baluška, František, Stefano Mancuso, Dieter Volkmann, and Peter W. Barlow. 2010. "Root Apex Transition Zone: A Signalling-Response Nexus in the Root." *Trends in Plant Science* 15 (7): 402–8. https://doi.org/10.1016/j.tplants.2010.04.007.

Bechtel, William. 2008. *Mental Mechanisms*. New York: Psychology Press.

Block, Ned. 1981. "Psychologism and Behaviorism." *The Philosophical Review* 90 (1): 5–43. https://doi.org/10.2307/2184371

Brenner, Eric D., Rainer Stahlberg, Stefano Mancuso, Jorge Vivanco, František Baluška, and Elizabeth Van Volkenburgh. 2006. "Plant Neurobiology: An Integrated View of Plant Signaling." *TRENDS in Plant Science* 11 (8): 413–19. https://doi.org/10.1016/j.tplants.2006.06.009

Burge, Tyler. 2010. *Origins of Objectivity*. Oxford: Oxford University Press.

Burge, Tyler. 2022. *Perception. First Form of Mind*. Oxford: Oxford University Press.

Cahill, James F., and Gordon G. McNickle. 2011. "The Behavioral Ecology of Nutrient Foraging by Plants." *Annual Review of Ecology, Evolution, and Systematics* 42: 289–311. https://doi.org/10.1146/annurev-ecolsys-102710-145006

Cahill, James F., Gordon G. McNickle, Joshua J. Haag, Eric G. Lamb, Samson M. Nyanumba, and Colleen Cassady St. Clair. 2010. "Plants Integrate Information About Nutrients and Neighbors." *Science* 328: 1657. https://doi.org/10.1126/science.1189736

Calvo, Paco. 2016. "The Philosophy of Plant Neurobiology: A Manifesto." *Synthese* 193: 1323–43. https://doi.org/10.1007/s11229-016-1040-1

Calvo, Paco, and Karl Friston. 2017. "Predicting Green: Really Radical (Plant) Predictive Processing." *Journal of the Royal Society Interface* 14: 20170096. https://doi.org/10.1098/rsif.2017.0096

Calvo, Paco, Monica Gagliano, Gustavo M. Souza, and Anthony Trewavas. 2020. "Plants Are Intelligent, Here's How." *Annals of Botany* 125: 11–28. https://doi.org/10.1093/aob/mcz155

Calvo, Paco, and Fred Keijzer. 2011. "Plants: Adaptive Behavior, Root-Brains, and Minimal Cognition." *Adaptive Behavior* 19 (3): 155–71. https://doi.org/10.1177/1059712311409446

Calvo, Paco, and Anthony Trewavas. 2020. "Physiology and the (Neuro)biology of Plant Behavior: A Farewell to Arms." *Trends in Plant Science* 25 (3): 214–16. https://doi.org/10.1016/j.tplants.2019.12.016

Chaiwanon, Juthamas, Wenfei Wang, Jia-Ying Zhu, Eunkyoo Oh, and Zhi-Yong Wang. 2016. "Information Integration and Communication in Plant Growth Regulation." *Cell* 164: 1257–68. https://doi.org/10.1016/j.cell.2016.01.044

Chamovitz, Daniel. 2012. *What a Plant Knows*. Oxford: Oneworld Publications.

Choi, Won-Gyu, Richard Hilleary, Sarah J. Swanson, Su-Hwa Kim, and Simon Gilroy. 2016. "Rapid, Long-Distance Electrical and Calcium Signaling in Plants." *Annual Reviews of Plant Biology* 67: 287–307. https://doi.org/10.1146/annurev-arplant-043015-112130

Churchland, Paul. 2012. *Plato's Camera*. Cambridge, MA: MIT Press.

Cvrčková, Fatima, Helena Lipavská, and Viktor Zárský. 2009. "Plant Intelligence." *Plant Signaling & Behavior* 4 (5): 394–9. https://doi.org/10.4161/psb.4.5.8276

Darwin, Charles. 1862. *On the Various Contrivances by Which British and Foreign Orchids Are Fertilised by Insects*. London: John Murray.

Davidson, Donald. 1982. "Rational Animals." *Dialectica* 36 (4): 317–27.

Dennett, Daniel. 2009. "Darwin's 'Strange Inversion of Reasoning'." *PNAS* 106 (1): 10061–5. https://doi.org/10.1073/pnas.0904433106.

Drew, M. C., and L. R. Saker. 1978. "Nutrient Supply and the Growth of the Seminal Root System in Barley." *Journal of Experimental Botany* 29 (109): 435–51.

Eliasmith, Chris. 2013. *How to Build a Brain*. Oxford: Oxford University Press.

Forterre, Yoël, Jan M. Skotheim, Jacques Dumais, and L. Mahadevan. 2005. "How the Venus Flytrap Snaps." *Nature* 433: 421–5. https://doi.org/10.1038/nature03185.

Gardner, Michael J., Katharine E. Hubbard, Carlos T. Hotta, Antony N. Dodd, and Alex A. R. Webb. 2006. "How Plants Tell the Time." *Biochemical Journal* 397: 15–24. https://doi.org/10.1042/BJ20060484.

Gladziejewski, Pawel, and Marcin Milkowski. 2017. "Structural Representations: Causally Relevant and Different from Detectors." *Biology & Philosophy* 32: 337–55. https://doi.org/10.1007/s10539-017-9562-6.

Grossberg, Stephen. 2021. *Conscious Mind, Resonant Brain*. Oxford: Oxford University Press.

Hodge, Angela. 2004. "The Plastic Plant: Root Responses to Heterogeneous Supplies of Nutrients." *New Phytologist* 162: 9–24. http://dx.doi.org/10.1111/j.1469-8137.2004.01015.x.

Hodge, Angela. 2009. "Root Decisions." *Plant, Cell and Environment* 32: 628–40. https://doi.org/10.1111/j.1365-3040.2008.01891.x.

Hodick, Dieter, and Andreas Sievers. 1989. "On the Mechanism of Trap Closure of Venus Flytrap (Dionea muscipula Ellis)." *Planta* 179: 32–42. https://doi.org/10.1104/pp.107.108241

Jones, Benji. 2023. "The Mystery of the Mimic Plant." *Vox*. https://www.vox.com/down-to-earth/2022/11/30/23473062/plant-mimicry-boquila-trifoliolata.

Kobayashi, Yasushi, and Detlef Weigel. 2007. "Move on Up, it's Time for Change - Mobile Signals Controlling Photoperiod-dependent Flowering." *Genes & Development* 21 (2371–2384). https://doi.org/10.1101/gad.1589007

Lee, Jonny, Miguel Segundo-Ortin, and Paco Calvo. 2023. "Decision Making in Plants: A Rooted Perspective." *Plants* 12: 1799. https://doi.org/10.3390/plants12091799.

Lindner, Ines. 2020. "Technik und Magie: Benjamin, Blossfeldt und 'Das Blumenwunder'." In *Internationales Jahrbuch für philosophische Anthropologie*, edited by Bruno Accarino, Jos de Mul, and Hans-Peter Krüger, 187–221. Berlin: De Gruyter. https://doi.org/10.1515/jbpa-2019-0011.

Lloyd, Dan. 1989. *Simple Minds*. Cambridge, MA: MIT Press.

Maher, Chauncey. 2017. *Plant Minds: A Philosophical Defense*. London: Routledge.

Mallatt, Jon, Michael R. Blatt, Andreas Draghun, David G. Robinson, and Lincoln Taiz. 2021. "Debunking a Myth: Plant Consciousness." *Protoplasma* 258: 459–76. https://doi.org/10.1007/s00709-020-01579-w.

Mancuso, Stefano, and Frantisek Baluška. 2017. "Plant Ocelli for Visually Guided Plant Behavior." *Trends in Plant Science* 22 (1): 5–6. https://doi.org/10.1016/j.tplants.2016.11.009.

Mancuso, Stefano, and Alessandra Viola. 2015. *Brilliant Green*. Washington, DC: Island Press.

Marr, David. 1982/2010. *Vision*. Cambridge, MA: MIT Press.

Massa, Gioia D., and Simon Gilroy. 2003. "Touch Modulates Gravity Sensing to Regulate the Growth of Primary Roots of *Arabidopsis thaliana*." *The Plant Journal* 33: 435–45. https://doi.org/10.1046/j.1365-313x.2003.01637.x.

Millikan, Ruth. 1984. *Language, Thought, and Other Biological Categories*. Cambridge, MA: MIT Press.

Millikan, Ruth. 2004. *Varieties of Meaning*. Cambridge, MA: MIT Press.

Millikan, Ruth. 2009. "Biosemantics." In *The Oxford Handbook of Philosophy of Mind*, edited by Brian McLaughlin, Ansgar Beckermann and Sven Walter, 394–406. Oxford: Oxford University Press. https://doi.org/10.1093/oxfordhb/9780199262618.003.0024.

Milner, A. David, and Melvyn A. Goodale. 2006. *The Visual Brain in Action*. 2nd ed. Oxford: Oxford University Press.

Morgan, Alex. 2018. "Mindless Accuracy: The Ubiquity of Content in Nature." *Synthese* 195: 5403–29. https://doi.org/10.1007/s11229-018-02011-w.

Moulton, Derek E., Hadrien Oliver, and Alain Goriely. 2020. "Multiscale Integration of Environmental Stimuli in Plant Tropisms Produces Complex Behaviors." *PNAS* 117 (51): 32226–37. https://doi.org/10.1073/pnas.2016025117.

Muthert, Lucius Wilhelminus Franciscus, Luigi Gennaro Izzo, Martijn van Zanten, and Giovanna Arrone. 2020. "Root Tropisms: Investigations on Earth and in Space to Unravel Plant Growth Direction." *Frontiers in Plant Science* 10: 1–22. https://doi.org/10.3389/fpls.2019.01807.

Nicholson, Charles C., James W. Bales, Joyce E. Palmer-Fortune, and Robert G. Nicholson. 2008. "Darwin's Bee Trap: The Kinetics of Catasetum, a New World Orchid." *Plant Signaling & Behavior* 3 (1): 19–23. https://doi.org/10.4161/psb.3.1.4980.

Ninkovic, Velemir, Dimitrije Markovic, and Merlin Rensing. 2021. "Plant Volatiles as Cues and Signals in Plant Communication." *Plant, Cell & Environment* 44: 1030–43. https://doi.org/10.1111/pce.13910.

Novoplansky, Ariel. 2019. "What Plant Roots Know?" *Seminars in Cell and Developmental Biology* 92: 126–33. https://doi.org/10.1016/j.semcdb.2019.03.009.

Orlandi, Nico. 2020. "Representing as Coordinating with Absence." In *What Are Mental Representations?*, edited by Joulia Smortchkova, Krzysztof Dolega and Tobias Schlicht, 101–34. Oxford: Oxford University Press. https://doi.org/10.1093/oso/9780190686673.003.0005.

Pautz, Adam. 2021. *Perception*. London: Routledge.

Piccinini, Gualtiero. 2020. *Neurocognitive Mechanisms*. Oxford: Oxford University Press.

Pierik, Ronald, and Mieke de Wit. 2014. "Shade Avoidance: Phytochrome Signaling and Aboveground Neighbour Detection Cues." *Journal of Experimental Botany*

65 (11): 2815–24. Shade avoidance: phytochrome signaling and aboveground neighbour detection cue.

Pierik, Ronald, and Christa Testerink. 2014. "The Art of Being Flexible: How to Escape from Shade, Salt, and Drought." *Plant Physiology* 166: 5–22. https://doi .org/10.1104/pp.114.239160.

Ramsey, William. 2007. *Representation Reconsidered*. Cambridge: Cambridge University Press.

Reid, Chris R., Simon Garnier, Madeleine Beekman, and Tanya Latty. 2015. "Information Integration and Multiattribute Decision Making in Non-Neural Organisms." *Animal Behaviour* 100: 44–50. http://dx.doi.org/10.1016/j .anbehav.2014.11.010.

Rescorla, Michael. 2013. "Millikan on Honeybee Navigation and Communication." In *Millikan and Her Critics*, edited by Justine Kingsbury, Dan Ryder and Kenneth Williford, 87–102. Malden, MA: Wiley-Blackwell. https://doi.org/10 .1002/9781118328118.ch4.

Rescorla, Michael. 2021. "Nicholas Shea, Representation in Cognitive Science (Review)." *Philosophical Review* 130 (1): 180–5. https://doi.org/10.1215 /00318108-8699643.

Robinson, David G., and Andreas Draguhn. 2021. "Plants Have Neither Synapses Nor a Nervous System." *Journal of Plant Physiology* 263: 153467. https://doi .org/10.1016/j.jplph.2021.153467.

Schulte, Peter. 2015. "Perceptual Representations: A Teleosemantic Answer to the Breadth-of-Application Problem." *Biology & Philosophy* 30: 119–36. https:// doi.org/10.1007/s10539-013-9390-2.

Schulte, Peter. 2023. *Mental Content*. Cambridge: Cambridge University Press.

Schulte, Peter. ms. *Borderlands of the Mental: Where and How Representation Begins*. Unpublished manuscript.

Sender, Ron, and Ron Milo. 2021. "The Distribution of Cellular Turnover in the Human Body." *Nature Medicine* 27: 45–8. https://doi.org/10.1038/s41591-020 -01182-9.

Shea, Nicholas. 2018. *Representation in Cognitive Science*. Oxford: Oxford University Press.

Sterelny, Kim. 1995. "Basic Minds." *Philosophical Perspectives* 9: 251–70. https:// doi.org/10.2307/2214221.

Tabata, Ryo, Kumiko Suminda, Tomoaki Yoshii, Kentaro Ohyama, Hidefumi Shinohara, and Yoshikatsu Matsubayasho. 2014. "Perception of Root-derived Peptides by Shoot LRR-RKs Mediates Systemic N-demand Signaling." *Science* 346 (6207): 343–6. https://doi.org/10.1126/science.1257800.

Taiz, Lincoln, Daniel Alkon, Andreas Draghun, Angus Murphy, Michael Blatt, Chris Hawes, Gerhard Thiel, and David G. Robinson. 2019. "Plants Neither Possess nor Require Consciousness." 24 (8): 677–87. https://doi.org/10.1016/j .tplants.2019.05.008.

Taiz, Lincoln, Daniel Alkon, Andreas Draguhn, Angus Murphy, Michael Blatt, Gerhard Thiel, and David G. Robinson. 2020. "Reply to Trewavas et al. and Calvo and Trewavas." *Trends in Plant Science* 25 (3): 218–20. https://doi.org /10.1016/j.tplants.2019.12.020.

Trewavas, Anthony. 2005. "Green Plants as Intelligent Organisms." *Trends in Plant Science* 10 (9): 413–19. https://doi.org/10.1016/j.tplants.2005.07.005.

Trewavas, Anthony. 2014. *Plant Behavior & Intelligence*. Oxford: Oxford University Press.

Wohlleben, Peter. 2016. *The Hidden Life of Trees*. Carlton: Black Inc.

Wu, Jianqiang, and Ian T. Baldwin. 2010. "New Insights into Plant Responses to the Attack of Insect Herbivores." *Annual Review of Genetics* 44: 1–24. https://doi.org/10.1146/annurev-genet-102209-163500.

Yokawa, Ken, Rossella Fasano, Tomoko Kageshini, and František Baluška. 2014. "Light as Stress Factor to Plant Roots - Case of Root Halotropism." *Frontiers in Plant Science* 5: 718. https://doi.org/10.3389/fpls.2014.00718.

Part 2

Sensation and Perception

4 Not All Sensory Systems Are Information Channels
Outliers from Plant Biology and Beyond

Todd Ganson

4.1 Introduction

This chapter is concerned with the defining characteristics of sensory systems, features of sensory systems that set them apart from other biological systems in nature. A guiding assumption of the discussion to follow is that biological sensory systems are to be defined in functional terms. In seeking a definition of sensory systems, we aim to uncover their evolutionary function—what sensory systems have been selected to do. A leading hypothesis is that sensory systems have evolved to channel information about an organism's environment. Following Dretske (1988, 56), we might define "an organism's sensory systems as channels for the receipt of information about their external environment" (see also Keeley 2002; Ganson 2018, 2021). As we shall see, Dretske's definition captures something important about how sensory systems in nature typically contribute to fitness. Nonetheless, it is a mistake to *define* sensory systems in informational terms: not all sensory systems are information channels.

It is often fruitful to think of biological functions as emerging in response to challenges or opportunities in an organism's environment. This chapter is focused on a challenge posed by the environment when environmental conditions vary in ways relevant to an organism's needs. Organisms are routinely faced with changes in how critical resources and threats are distributed temporally and spatially, and these fluctuations in environmental conditions often impact whether a given behavior achieves its selected-for purpose. Consequently, organisms confront a coordination problem: they need to match their behavioral outputs with environmental conditions conducive to the biological success of the outputs (Godfrey-Smith 1996; Sterelny 2003). Many sensory systems in nature are plausibly interpreted as solutions to this coordination problem, henceforth referred to as *the problem of ecological complexity*.

The problem of ecological complexity is ubiquitous in nature. Think, for example, of a prey's need to coordinate its flight behavior with the presence of a predator. Suppose an organism's flight response is an adaptation

DOI: 10.4324/9781003393375-7

Table 4.1 Coordination of flight response with presence of a predator

	predator present	*predator absent*
flight response	hit (coordination)	false alarm (failure)
no flight response	miss (failure)	correct rejection (coordination)

to selective pressure exerted by predation. The defensive maneuver contributes to fitness insofar as it allows the organism to escape from the presence of a predator. The challenge facing the prey is to coordinate their output (flight response) with the environmental condition pertinent to the biological success of the output (presence of a predator). Stated in the terminology of signal detection theory (McNicol 1972), successful coordination takes the form of *hits* (flight response when a predator is present) and *correct rejections* (no flight response when predators are absent); failures of coordination manifest as either *misses* (no flight response when a predator is present) or *false alarms* (flight response when predators are absent).

Sensory systems are useful in confronting the problem of ecological complexity because they can help make a positive impact on the rate at which organisms achieve coordination between their behavior and suitable environmental conditions. For example, some marine annelids react to sudden decreases in illumination with a flight response (Ayers et al. 2018). The worms are hard-wired to produce their evasive maneuver in response to sudden onset of darkness because this adaptation helps them evade shadow-casting predators looming above. This behavioral strategy deploys a light-sensitive sensory system in the interest of promoting the success rate of the flight response. Borrowing once again from signal detection theory, we can distinguish several measures of success rate.

- Accuracy: the sum of hits and correct rejections divided by the total number of observed trials
- Hit rate: the number of hits divided by the sum of hits and misses (left column)
- Precision: the number of hits divided by the sum of hits and false alarms (upper row)

Which type of success is important varies across different outputs and different ecological contexts. We expect the worm's behavioral strategy to prioritize hit rate over, say, precision because the costs for the worm associated with misses will typically outweigh the costs associated with false alarms.

The standard way organisms address the problem of ecological complexity is by conditioning the output in question on stimulus conditions: the organism produces the output conditional upon receipt of a specific stimulus. (Throughout what follows, the term "stimulus" refers to the *proximal stimulus* which initiates receptor signaling.) The organism is set up to be set off by the stimulus and thereby makes use of the stimulus *as a stimulus or catalyst*. For example, the aforementioned marine annelids succeed in promoting the hit rate of their flight behavior by conditioning the behavior on sudden onset of darkness. The worms are hard-wired to produce the output in response to the stimulus, and they thereby exploit the stimulus as a stimulus. When we say that an organism is utilizing a stimulus as a stimulus, we are talking about a *causal role* of the stimulus. The stimulus has been recruited through natural selection to serve as a triggering cause of the behavior. The organism may also be exploiting the stimulus for information it carries, but use of a stimulus as a stimulus does not logically entail use of a stimulus for information.

Why might conditioning behavior on a stimulus help the organism coordinate its behavior with suitable environmental conditions? Proponents of the informational approach to sensory systems have a straightforward answer: conditioning the behavior on a stimulus condition brings alignment between the behavior and suitable environmental conditions because the stimulus *correlates* with those environmental conditions. For example, the worms boost the hit rate of their evasive maneuver by conditioning the response on a stimulus condition that *raises the probability* of presence of a predator. The worms are making use of *correlational information* carried by the stimulus. A stimulus S carries correlational information about an environmental condition E if and only if S changes the objective probability of E, i.e. $P(E|S) \neq P(E)$, and this correlation between S and E is non-spurious—it persists for a reason (Scarantino 2013; Stegmann 2015; Shea 2018). The worms manage to address the problem of ecological complexity by conditioning their flight response on a stimulus that carries correlational information about the threat of predation. They have evolved to condition flight behavior on this stimulus condition *because the stimulus carries correlational information about the threat of predation.*

A plausible precondition for a sensory system to count as making use of information is that the sensory system has the kind of structural complexity characteristic of an information channel. In its canonical form, an information channel possesses a sender-receiver structure with signals conveyed from the sender to the receiver (Shannon 1948; Lewis 1969; Skyrms 1996, 2010). A sender in a sender-receiver configuration (signaling system) is something capable of receiving a stimulus as input and generating a signal as output, while a receiver is something capable of receiving the signal as input and generating an activity as output. Sensory systems satisfy this requirement by decomposing into structurally distinct receptor and effector elements. The receptor unit moves from receipt of a stimulus as input to production of a sensory signal as output, and the effector unit moves from receipt of the signal as input to production of an active response to the signal. For example, our worms have photoreceptors which signal in response to the stimulus (sudden onset of darkness), and they have a motor unit that is triggered by receipt of the signal.

Sensory systems with this kind of internal complexity are well suited to address the problem of ecological complexity, which requires an organism to coordinate a specific output O with a specific state of the environment E. A signaling system can promote this sort of coordination between O and E by exploiting a stimulus condition S. The signaling system makes use of S in two ways. First, the system exploits S as a stimulus or catalyst. The system is set up to be set off by S, with S triggering receptor signaling and the receptor-issued signal, in turn, triggering the receiver's response O. In this way, the sender-receiver configuration allows an organism to condition O on S. Second, the system exploits the stimulus as a bearer of information. Suppose that S carries correlational information about E. More specifically, suppose S raises the probability of E, i.e. $P(E|S) > P(E)$. By conditioning O on S, the organism can thereby increase its hit rate (coordination of O with E when E obtains).

These preliminary remarks set the stage for understanding how channeling information can help an organism address the problem of ecological complexity. However, there is a significant obstacle that remains. According to proponents of information-based definitions of sensory systems, sensory signals are recruited to *inform* the organism's behavior. The signal is not just a causal intermediary between stimulus and response; it has the *function of conveying information* about the environment. In order to establish that the sensory system is indeed functioning as an information channel, we need to show that the sensory signal's job does not reduce to serving as a causal intermediary between stimulus and response (Ramsey 2007; Ganson 2020). We need to show that the sensory system's functional role goes beyond its role in conditioning the response upon receipt of the stimulus. Clearly, it will not be enough to show that the sensory signal carries correlational information about the environment. A sensory signal can come to carry information about an environmental condition just by virtue of being reliably triggered by the environmental condition. We do not need to acknowledge a role beyond that of causal intermediary to make sense of the fact that the sensory signal carries information about the environment.

We can hope to identify a role for information in sensory signaling by getting clear about the behavioral strategy that is being deployed. Once we know what purpose is achieved by conditioning the output on the stimulus, we can ask whether information carried by the sensory signal helps to *explain the success of the behavioral strategy*. (For the idea that the relevant explanandum is behavioral success, see Shea 2018; Schulte 2019; Ganson 2020.) Our focus here is on sensory systems that have been selected to promote coordination between the output and suitable environmental conditions. Our question, then, is whether information carried by the sensory signal plays an indispensable role in promoting coordination. The proposal developed in §2 is that we can get the evidence we are looking for by making information-specific interventions on the stimulus conditions. The goal of these manipulations is to show that information carried by the sensory signal is an independent variable influencing the success rate of the organism's response to the stimulus. When information is an additional contributing factor in confronting the problem of ecological complexity, the sensory signal's role does not reduce to serving as a causal intermediary. The sensory signal's job goes beyond *triggering* the response; it also has the role of *informing* the response.

There are, however, reasons to think that some sensory systems in nature address the problem of ecological complexity just by exploiting the stimulus as a stimulus. In these cases being set up to be set off by the stimulus suffices to promote coordination between the output and suitable environmental conditions, so there is no explanatory role for information to play in achieving behavioral success. Sensory systems that fit this description serve as counterexamples to informational approaches to sensory systems. The counterexamples surveyed in §3 have significance beyond the question of how to define sensory systems. They are also relevant to theorizing about the cognitive powers of creatures like plants. As we shall see in §4, we have to take care when inferring that a creature has information-processing capabilities on the basis of its possession of sensory systems. We have to determine whether the creature has sensory systems recruited by natural selection to play the role of information channels.

4.2 How Information Can Make a Difference

The focus of this chapter is on sensory systems that have evolved to tackle the problem of ecological complexity. The function of these sensory systems is to coordinate the organism's behavioral output with environmental conditions conducive to the success of the output. We are assuming that the sensory system contributes to coordination by allowing the organism to condition the output on stimulus conditions. The organism's sensory system has evolved to exploit the stimulus as a stimulus: the receptor unit moves from receipt of the stimulus to production of a sensory signal and the effector unit moves from the signal to production of the behavioral output. The question at hand is whether the sensory system has the additional job of informing the response. On the supposition that information channeling is indeed integral to the way a sensory system promotes the success rate of the organism's behavior, we would predict that manipulating the correlational information carried by the sensory signal will affect the success rate. By performing information-specific interventions, we can hope to isolate a role for information in confronting the problem of ecological complexity.

Consider again the defensive maneuver exhibited by our marine annelids in response to sudden decreases in illumination. Let us assume that the worms have a successful strategy for confronting the problem of ecological complexity: they succeed in promoting the hit rate of their flight behavior by conditioning the behavior on receipt of a specific stimulus (sudden onset of darkness). We are taking for granted that the strategy involves exploiting the stimulus *as a stimulus*, and we want to know whether the worms are also making use of the stimulus *as a bearer of correlational information*. The first question to ask is whether the stimulus carries correlational information about the presence of a predator: Does the stimulus raise the objective probability that a shadow-casting predator looms above? Suppose the answer is *yes*. Our next question is whether the worms condition their flight response on sudden onset of darkness *because the stimulus carries correlational information about the presence of a predator*. Does their behavioral strategy depend crucially on the fact that the stimulus raises the probability of the presence of a predator? If the answer is again *yes*, we predict that intervening on the correlation between sudden onset of darkness and presence of a predator will affect the success rate of the flight behavior.

We can experimentally isolate the contribution information makes to the success of the worm's behavioral strategy by manipulating the correlation between sudden onset of darkness and presence of a predator in the local environment of the test subjects, holding fixed the natural tendency of these worms to be set off by absence of light. The independent variable of our experiment is the extent to which sudden onset of darkness raises the probability of the presence of a predator in the creature's environment. The dependent variable is the hit rate of the withdrawal response. We are able to extricate the stimulus's role as bearer of information from its role as stimulus by making information-specific interventions on the stimulus condition, leaving the stimulus's role as a stimulus unaltered. If the interventions affect the success rate of the response, we have evidence the worms are making use of the stimulus for information it carries about the presence of a predator.

We have assumed from the outset that our marine annelids confront the problem of ecological complexity by conditioning their flight response upon receipt of a specific stimulus. The question we raised is whether this behavioral strategy *reduces* to utilizing the stimulus as a stimulus. Can we exhaustively analyze the success of the strategy in terms of the stimulus's causal influence on the worm's behavior? The alternative hypothesis is that the stimulus's causal influence does not by itself suffice to promote coordination between the worm's flight response and presence of a predator; the stimulus condition must also carry information about the presence of a predator. We can hope to shed light on this question by making information-specific interventions on the stimulus condition. If these interventions affect the hit rate of the response, we have evidence that correlational information is an independent variable affecting whether the worms achieve coordination between their flight response and the presence of a predator. We have evidence that information carried by the stimulus is making a contribution over and above the stimulus's causal influence on behavior.

Advocates of information-based theories of sensory systems insist that what is distinctive about sensory systems in nature is their functional role in informing the organism's behavior. They take it to be an essential or defining characteristic of sensory systems that they have been recruited by selective forces to channel information. One way to motivate this view is to emphasize how integral information is to solving the problem of ecological complexity. The proponent of an informational view of sensory systems can reasonably ask why conditioning an output on a stimulus condition would help to promote coordination between the output and a suitable state of the environment *unless the stimulus correlates with that state of the environment*. Conditioning the behavior on the stimulus is supposed to be a strategy for improving the success rate of the behavior, and it is difficult to understand how this strategy can be successful unless the stimulus increases the likelihood that the environmental state obtains.

4.3 Not All Sensory Systems Are Information Channels

The goal up to this point has been to identify a role for information in confronting the problem of ecological complexity. The preliminary conclusion reached so far is that information channeled by a sensory system can make a distinctive contribution to coordination over and above the causal role of the sensory system in moving from stimulus to response. Information carried by the sensory signal can be an independent variable impacting the rate at which the organism successfully aligns its behavioral outputs with appropriate environmental conditions. It does not follow, however, that channeling information is an essential feature of sensory systems in nature. The aim in this section is to raise doubts about the idea that we should define sensory systems in informational terms. Sometimes a system's causal role is sufficient to account for success in addressing the problem of ecological complexity. There is no role for information to play in achieving coordination between outputs and suitable environmental conditions.

The cases that make trouble for information-based accounts of sensory systems are cases where the stimulus condition is itself the state of the world that makes a difference to the success of the organism's output. The organism needs to align its output with the stimulus itself, not with something that correlates with the stimulus. Here are some examples:

- The phototrophic consortium *Chlorochromatium aggregatum* displays a negative tactic response to absence of light (Fröstl and Overmann 1998). As a phototroph, this bacterial consortium aims to remain in light and avoid absence of light. Absence of light is both the stimulus that triggers the negative tactic response and the environmental condition of importance to the microbial community, the state of the world to be avoided. The goal is to coordinate the response (repulsion) with the stimulus condition (absence of light), not with something that correlates with the stimulus. Accordingly, biological success does not require making use of the stimulus as a bearer of information about something further. It is enough to be set up to be set off by the stimulus condition.
- In the proboscis extension reflex, honey bees protract their tubular tongue in response to stimulation of antennal sugar receptors (Giurfa and Sandoz 2012). This feeding behavior has evolved to coordinate the output with the stimulus itself. The sugars that trigger the response are themselves conducive to the success of the response.
- Common water fleas of the genus *Daphnia* exhibit a negative phototactic response to harmful UV radiation (Tollrian et al. 2001). The high-energy radiation is not serving as a bearer of information about something further. Exposure to high-energy radiation is important because the exposure is itself dangerous.

In each of these examples, the organism (or consortium) needs to coordinate its output with the stimulus condition itself, and it solves the problem just by exploiting the stimulus as a stimulus. The strategy of being set up to be set off by the stimulus is sufficient to address the problem of ecological complexity because the stimulus is the state of the environment that makes a difference to the success of the output.

Consider one last time the marine annelids that condition their avoidance behavior on absence of light. Like these worms, the phototrophic consortium *Chlorochromatium aggregatum* is also set up to be set off by absence of light. Nonetheless, the two behavioral strategies are quite different in character. The worms flee in response to absence of light because this stimulus condition correlates with shadow-casting predators looming above. Their behavioral strategy depends on channeling information: it is not enough for the worms to be wired-up to flee in response to absence of light; correlational information carried by the sensory signal is also integral to coordinating the flight response with presence of a predator. The information conveyed by the sensory system is an independent variable contributing to behavioral success. The bacterial consortium, by contrast, achieves successful coordination just by being set up to be set off by the stimulus condition. Since the consortium needs to flee from absence of light itself, all it needs to do is exploit this stimulus condition as a stimulus.

In both cases the sensory system produces a sensory signal that carries correlational information about the environment. But carrying correlational information is not the same thing as having the functional role of carrying correlational information. We need to draw a distinction between two ways a sensory signal might come to carry information:

- The sensory signal has been recruited by natural selection to serve as a causal intermediary between stimulus and response because the signal carries correlational information about the environment.
- The sensory signal carries correlational information about the environment as a consequence of the fact that the signal has been recruited by natural selection to serve as a causal intermediary between stimulus and response.

Only in the former case should we say that the sensory signal has the *function* of informing the response. In order for the signal to have the job of conveying information, the signal's carrying information has to be integral to why it was recruited to elicit the response. For example, the worm's sensory signal has been recruited to trigger flight behavior because it carries information about the presence of a predator. In the second type of case, by contrast, the sensory signal was recruited to serve as a causal intermediary between stimulus and response. It has evolved to be reliably triggered by the stimulus condition. Consequently, the sensory signal raises the probability that the stimulus condition is present, and thereby carries information about the stimulus condition. The consortium's avoidance behavior illustrates this kind of case.

Why say that the consortium's sensory signal is limited to the role of serving as causal intermediary between stimulus and response? Because being set up to be set off by the stimulus is sufficient to address the problem of ecological complexity. The consortium needs to align its avoidance behavior with absence of light, so it suffices for the consortium to condition its avoidance behavior on absence of light. The consortium is able to achieve successful coordination just by exploiting the stimulus as a stimulus. Since the consortium's sensory processes are reliably triggered by absence of light, they end up carrying correlational information about absence of light. But information is explanatorily idle here. The behavioral strategy *reduces to* making use of the stimulus as a stimulus.

In order for correlational information to play an explanatory role in achieving successful coordination, it needs to make a difference to whether coordination is achieved. The correlational information needs to be an independent variable influencing the biological success of the output. Otherwise success can be exhaustively analyzed in terms of exploiting the stimulus as a stimulus. We establish this additional use of the stimulus by holding fixed the organism's (or, in the case at hand, consortium's) use of the stimulus as a stimulus. When information-specific interventions affect the success of the behavioral strategy, we have evidence that the strategy is information dependent. This kind of manipulation is straightforwardly ruled out in cases like the consortium's use of absence of light. We cannot intervene in the correlation between stimulus and absence of light because the stimulus *is* absence of light. A stimulus's correlation with itself is not something susceptible to experimental manipulation. In the absence of this additional use of the stimulus for information, the behavioral strategy reduces to exploiting the stimulus as a stimulus.

4.4 The Case of Plants

The problem of ecological complexity is not restricted to cases where the output options are *behavioral*; it also arises when a creature has *developmental* options. Plants provide a helpful illustration. There is considerable variation in how critical resources for plants are distributed temporally and spatially. Resources like light, water, and mineral nutrients are unevenly distributed over space and time. As sessile organisms, plants are unable to change location in order to address their needs. They do, however, have recourse to adjusting developmental outputs to fit with environmental conditions. They can do so by deploying sensory systems that allow them to condition their developmental outputs on stimulus conditions. This strategy of exploiting stimuli as stimuli can take both of the forms outlined above. In some cases plants condition an output on a stimulus in order to make additional use of the stimulus as a bearer of correlational information about the environment; in other cases plants are able to meet the challenge posed by variation in environmental conditions just by exploiting the stimulus as a stimulus. Cases of the latter sort are especially relevant in the context of thinking about the cognitive capacities of plants: they highlight the need to proceed with caution when attributing information-gathering powers to plants.

We begin with an example of a plant making use of a stimulus for information it carries. Plants in the family Orobanchaceae are parasitic and need to coordinate germination with the presence of a suitable host plant. Their strategy involves making germination conditional upon receipt of a specific type of hormone (strigolactones), which plants discharge into the surrounding soil (Conn et al. 2015). This strategy improves the hit rate of

Table 4.2 Coordination of germination with host presence

	suitable host present	suitable host absent
germination	hit (coordination)	false alarm (failure)
no germination	miss (failure)	correct rejection (coordination)

the response because these strigolactones raise the probability that a suitable host plant is present. The conditional probability of a suitable host plant being present, given the stimulus, is higher than the unconditional probability that a suitable host plant is present. Accordingly, the strategy of conditioning germination on receipt of the stimulus can increase the rate at which these parasites achieve coordination as compared to the strategy of acting regardless of stimulus conditions. By conditioning germination on receipt of the stimulus, the plant is not only making use of the stimulus as a stimulus; it is also making use of the stimulus for information it carries.

Next consider a plant's ability to regulate the spatial direction of growth. Given the largely random manner in which goods are distributed, one strategic option would be for plants to randomize the directions in which shoots and roots grow. In plant tropisms we find an alternative to randomization: plants adopt the strategy of conditioning growth patterns on stimulus conditions. Phototropism involves adjusting direction of growth in response to light reaching the plant. In positive phototropism plants orient growth in the direction of greater quantities of light. This strategy involves exploiting light as a stimulus. Equipped with photoreceptors, plants are set up to be set off by incident light. Photoreceptor stimulation triggers a redistribution of auxin, which, in turn, alters the orientation of the organ stimulated.

There is a broad consensus that plant phototropism is a sensory process. Plants are relying on a sensory system in regulating directional orientation. It does not follow that these shifts in orientation are to be explained in information-theoretic terms. Consider the possibility that plants have the goal of adjusting to current conditions along the lines of the pupillary light reflex. The pupillary light reflex functions to adjust pupil aperture with the aim of optimizing light capture. The stimulus of light intensity at the eyes is not being used as a bearer of information about something further; light intensity at the eyes is itself the biologically important aspect of the environment with which the output (pupil size) needs to coordinate. Of course, the pupillary light reflex operates on a shorter time frame than plant phototropic responses. Nonetheless, it is plausible that the objective is broadly the same: to bring responses in line with incident light, making appropriate adjustments to light intensity as changes occur. On this way of thinking about plant phototropism, the goal is to get outputs to coordinate with the triggering stimulus conditions themselves, not with something that correlates with light intensity. Hence, success does not depend on exploiting the stimulus for information; it is enough to exploit the stimulus as a stimulus.

The main worry about assimilating plant phototropism to the pupillary light reflex is that the outputs of the two processes can be rather different in character. Directional growth is not always reversible in the way constrictions and dilations of the pupil are. In favoring one direction of growth over another, plants are sometimes making a longer-term commitment. Accordingly, phototropism sometimes exploits the stimulus for information about more enduring aspects of the plant's surroundings. The idea is that plants are sampling stimulus conditions in an effort to be optimally, or at least satisfactorily, oriented relative to reliable light sources going forward. The stimulus is exploited for information it carries about the direction of more plentiful light sources. To be clear, the suggestion is not that sensory systems channel information about the future; the plant's sensory processes convey information about present lighting conditions. However, some aspects of an environment are more likely to persist than others, and sensory systems can be used to garner information about these relatively permanent aspects of the environment. Think of a bird relying on landmarks in its caching behavior.

Suppose we embrace this line of thought, and take irreversibility to be evidence that plants utilize light for information. By the same token, we should treat reversibility as evidence that the process is akin to the pupillary light reflex. As it turns out, much of the phototropic activity we find in plants fits this model well. Think of the familiar oscillations of the sunflower or the robust reversibility manifest in the phototropic responses of potato shoots (Vinterhalter et al. 2016). It is a mistake, then, to infer that a plant is utilizing information from the fact that it exhibits phototropic bending. Phototropism is not inherently an information-utilizing process.

4.5 Conclusion

This chapter has focused on sensory systems that have evolved to confront the problem of ecological complexity, the problem of coordinating a behavioral or developmental output with environmental circumstances conducive to the biological success of the output. Sensory systems can contribute to fitness by positively impacting the rate at which an organism achieves coordination between the output and suitable environmental conditions. The strategy involves conditioning the output on stimulus conditions. The sensory system is a signaling system that allows the organism to produce the output conditional upon receipt of the stimulus: the sender (receptor unit) moves from receipt of the stimulus to production of a sensory signal and the receiver (effector unit) moves from the signal to production of the output. In this way the organism is set up to be set off by the stimulus and thereby exploits the stimulus as a stimulus or catalyst. The question is whether this strategy also involves exploiting the stimulus for information it carries.

We want to know whether a sensory system confronts the problem of ecological complexity by channeling information about the environment. How might we establish that a sensory system is functioning as an information channel? The first thing to note is that a sensory signal can come to carry information about the environment just in virtue of being reliably triggered by environmental conditions. Accordingly, we ought to distinguish two possibilities:

- The sensory signal has been recruited by natural selection to serve as a causal intermediary between stimulus and response because it carries correlational information about the environment.
- The sensory signal carries correlational information about the environment as a consequence of the fact that the signal has been recruited by natural selection to serve as a causal intermediary between stimulus and response.

A sensory system with the function of channeling information is more than just a causal intermediary between stimulus and response. When a sensory system has been recruited to serve as an information channel in response to the problem of ecological complexity, information carried by the signal helps explain why the system is successful in confronting the challenge. The signal was recruited to serve as a causal intermediary between stimulus and response because it carries information about the environment; it doesn't carry information about the environment because it was recruited to be a causal intermediary.

On the supposition that a sensory system addresses the problem of ecological complexity by channeling information, information carried by the sensory signal makes a difference to the success rate of the response. Information is an independent variable influencing the success rate. Conditioning the response on the stimulus does not suffice for coordinating the response with suitable environmental conditions; the sensory system must also channel information carried by the stimulus. We can hope to uncover a crucial role for information by making information-specific interventions on the stimulus, leaving the stimulus's role as a stimulus unaltered. If these information-specific manipulations affect the success rate of the response, we have evidence that the sensory system is functioning as an information channel. The sensory system's role goes beyond just serving as a causal intermediary between stimulus and response; the sensory system has the job of informing the organism's response.

Once we grant that the job of channeling information goes beyond serving as a causal intermediary between stimulus and response, we have to acknowledge the possibility that some sensory systems do not count as information channels. We have to allow for the possibility that a sensory system's success in confronting the problem of ecological complexity is exhaustively analyzable in terms of its causal role in producing the output conditional upon receipt of the stimulus. For example, information seems to be explanatorily superfluous when an organism achieves coordination between an output and a suitable environmental condition by exploiting the environmental condition itself as a stimulus. Being set up to be set off by the stimulus suffices for promoting coordination because the stimulus is the state of the world with which the organism needs to coordinate its output. There is no explanatory role for information to play in confronting the problem of ecological complexity, so the strategy reduces to exploiting the stimulus as a stimulus.

We began with the question of how sensory systems differ from other biological adaptations. The argument of this chapter speaks against defining sensory systems in informational terms. We should not conclude, however, that sensory systems are not signaling systems (Ganson 2018). The concept of a signaling system has application to a broad range of biological systems in nature. It applies not only to intraorganismal signaling of the sort carried out by sensory systems but also to interorganismal signaling manifest in animal communication and to intercellular signaling in cell communication. In the context of animal communication, plenty of theorists reject the assumption that signaling systems should be defined in informational terms (Artiga 2021; Dawkins and Krebs 1978; Krebs and Dawkins 1984; Owren, Rendall, and Ryan 2010; Rendall, Owren, and Ryan 2009; Scott-Phillips 2008, 2010). In this regard they are sympathetic with Maynard Smith and Harper's (2003, 3) influential definition of a signal "as any act or structure which alters the behavior of other organisms, which evolved because of that effect, and which is effective because the receiver's response has also evolved." Here the signal in a communication channel is defined by its causal role rather than by its role as a bearer of information. Presumably this approach to signaling can be adapted to fit the case of sensory signaling. The argument of this chapter recommends an approach to sensory signaling along these lines.

Acknowledgments

Thanks to Marc Artiga, Brian Keeley, Caitlin Mace, Peter Schulte and an anonymous referee for valuable comments and suggestions on previous versions of this chapter. Thanks also to the editors of this volume, and especially Peter Schulte. Schulte, Artiga and Keeley have all had a significant impact on my thinking about sensory cognition through their various publications and correspondences. Finally, thanks to audience members at several conferences where I presented material in this chapter, including sessions at the Central APA, Pacific APA and the conference *Green Intelligence?—Debating Plant Cognition.*

References

Artiga, Marc. 2021. "Signals are Minimal Causes." *Synthese (Dordrecht)* 198 (9): 8581–99. https://doi.org/10.1007/s11229-020-02589-0

Ayers, Thomas, Hisao Tsukamoto, Martin Guhmann, Vinoth Babu Veedin Rajan, and Kristin Tessmar-Raible. 2018. "A Go-Type Opsin Mediates the Shadow Reflex in the Annelid Platynereis Dumerilii." *BMC Biology* 16 (1): 41. https://doi.org/10.1186/s12915-018-0505-8.

Conn, Caitlin E., Rohan Bythell-Douglas, Drexel Neumann, Satoko Yoshida, Bryan Whittington, James H. Westwood, Ken Shirasu, Charles S. Bond, Kelly A. Dyer, and David C. Nelson. 2015. "Convergent Evolution of Strigolactone Perception Enabled Host Detection in Parasitic Plants." *Science (American Association for the Advancement of Science)* 349 (6247): 540–43. https://doi.org/10.1126/science.aab1140.

Dawkins, R., and J. Krebs. 1978. "Animal Signals: Information or Manipulation?" In *Behavioral Ecology: An Evolutionary Approach,* edited by John Krebs and Nicholas B. Davies, 282–309. Oxford: Blackwell Scientific.

Dretske, Fred I. 1988. *Explaining Behavior: Reasons in a World of Causes.* Cambridge, MA: MIT Press. https://doi.org/10.7551/mitpress/2927.001.0001.

Fröstl, J. M., and J. Overmann. 1998. "Physiology and Tactic Response of the Phototrophic Consortium Chlorochromatium Aggregatum." *Archives of Microbiology* 169 (2): 129–35. https://doi.org/10.1007/s002030050552.

Ganson, Todd. 2018. "The Senses as Signalling Systems." *Australasian Journal of Philosophy* 96 (3): 519–31. https://doi.org/10.1080/00048402.2017.1381749.

Ganson, Todd. 2020. "A Role for Representations in Inflexible Behavior." *Biology & Philosophy* 35 (4). https://doi.org/10.1007/s10539-020-09756-0.

Ganson, Todd. 2021. "An Alternative to the Causal Theory of Perception." *Australasian Journal of Philosophy* 99 (4): 683–95. https://doi.org/10.1080/00048402.2020.1836008.

Giurfa, Martin, and Jean-Christophe Sandoz. 2012. "Invertebrate Learning and Memory: Fifty Years of Olfactory Conditioning of the Proboscis Extension Response in Honeybees." *Learning & Memory* 19: 54–66. https://doi.org/10.1101/lm.024711.111.

Godfrey-Smith, Peter. 1996. *Complexity and the Function of Mind in Nature.* West Nyack: Cambridge University Press. https://doi.org/10.1017/CBO9781139172714.

Keeley, Brian L. 2002. "Making Sense of the Senses: Individuating Modalities in Humans and Other Animals." *The Journal of Philosophy* 99 (1): 5–28. https://doi.org/10.2307/3655759.

Krebs, John, and Richard Dawkins. 1984. "Animal Signals: Mind-Reading and Manipulation." In *Behavioural Ecology: An Evolutionary Approach,* edited by John Krebs and Nicholas B. Davies, 380–402. Oxford: Blackwell Scientific.

Lewis, David K. 1969. *Convention: A Philosophical Study.* Cambridge: Harvard University Press.

Maynard Smith, John, and David Harper. 2003. *Animal Signals.* Oxford: Oxford University Press.

McNicol, Don. 1972. *A Primer of Signal Detection Theory*. Mahwah: Psychology Press. doi:10.4324/9781410611949.

Owren, Michael J., Drew Rendall, and Michael J. Ryan. 2010. "Redefining Animal Signaling: Influence Versus Information in Communication." *Biology & Philosophy* 25 (5): 755–80.

Ramsey, William M. 2007. *Representation Reconsidered*. Cambridge, GBR: Cambridge University Press. https://doi.org/10.4324/9781410611949.

Rendall, Drew, Michael J. Owren, and Michael J. Ryan. 2009. "What Do Animal Signals Mean?" *Animal Behavior* 78 (2): 233–40. https://doi.org/10.1016/j.anbehav.2009.06.007.

Scarantino, Andrea. 2013. "Animal Communication as Information–Mediated Influence." In *Animal Communication Theory*, edited by Ulrich E. Stegmann, 63–88. Cambridge: Cambridge University Press. https://doi.org/10.1017/CBO9781139003551.005.

Schulte, Peter. 2019. "Challenging Liberal Representationalism: A Reply to Artiga." *Dialectica* 73 (3): 331–48. https://doi.org/10.1111/1746-8361.12275.

Scott-Phillips, T. C. 2008. "Defining Biological Communication." *Journal of Evolutionary Biology* 21 (2): 387–95. https://doi.org/10.1111/j.1420-9101.2007.01497.x.

Scott-Phillips, T. C. 2010. "Animal Communication: Insights from linguistic Pragmatics." *Animal Behavior* 79 (1): e1–e4. http://dx.doi.org/10.1016/j.anbehav.2009.10.013.

Shannon, Claude. 1948. "A Mathematical Theory of Communication." *The Bell System Mathematical Journal* 27 (3): 379–423, 623–56. https://doi.org/10.1002/j.1538-7305.1948.tb01338.x.

Shea, Nicholas. 2018. *Representation in Cognitive Science*. Oxford: Oxford University Press. https://doi.org/10.1093/oso/9780198812883.001.0001.

Skyrms, Brian. 1996. *Evolution of the Social Contract*. West Nyack: Cambridge University Press. https://doi.org/10.1017/CBO9781139924825.

Skyrms, Brian. 2010. *Signals: Evolution, Learning, and Information*. Oxford: Oxford University Press. https://doi.org/10.1093/acprof:oso/9780199580828.001.0001.

Stegmann, Ulrich E. 2013. *Animal Communication Theory: Information and Influence*. Edited by Ulrich E. Stegmann. New York: Cambridge University Press. https://doi.org/10.1017/CBO9781139003551.

Stegmann, Ulrich E. 2015. "Prospects for Probabilistic Theories of Natural Information." *Erkenntnis* 80 (4): 869–93. https://doi.org/10.1007/s10670-014-9679-9.

Sterelny, K. 2003. *Thought in a Hostile World: The Evolution of Human Cognition*. Oxford: Oxford University Press.

Tollrian, Ralph, Stephan C. Rhode, and Markus Pawlowski. 2001. "The Impact of Ultraviolet Radiation on the Vertical Distribution of Zooplankton of the Genus Daphnia." *Nature* (London) 412 (6842): 69–72. https://doi.org/10.1038/35083567.

Vinterhalter, D., J. Savić, M. Stanišić, Ž. Jovanović, and B. Vinterhalter. 2016. "Interaction with Gravitropism, Reversibility and Lateral Movements of Phototropically Stimulated Potato Shoots." *Journal of Plant Research* 129 (4): 759–70. https://doi.org/10.1007/s10265-016-0821-4.

5 Plants Sense. But Only Animals Perceive

Mohan Matthen

5.1 Introduction

All living things—plants, bacteria and other non-animals included—adapt their behaviour to their surroundings. To do this, they have organs that are sensitive to environmental conditions. In short, all living things have sensory capacities. The question for philosophers of perception is this: How do these capacities relate to reality?

Now, many philosophers are inclined to deny—indeed, to deny emphatically—that plants could have *cognitive* capacities or be conscious. So, the story goes, they do not mediate response by coming to know it or by being conscious of it. These assertions typically do not take into account what we know about plants. Michael Tye (2021), for example, writes, entirely without citation, that the inner states of "tropistic organisms" such as plants:

> are surely not phenomenal. There is nothing it is like to be a Venus Fly Trap or a Morning-Glory. [...] The behavior of plants is inflexible. It is genetically determined and, therefore, not modifiable by learning [...] Plants do not learn from experience [...] are not subject to any qualia. Nothing that goes on inside them is poised to make a direct difference to what they believe or desire, since they have no beliefs or desires.
>
> (ibid., §12)

The problem here is not so much that Tye's conclusions are false. It is rather that, as we shall see, they are oblivious to relevant distinctions (some arrived at through decades, or even centuries, of empirical investigation) and are, as such, misleading or worse.[1]

In the current state of scientific knowledge, it is becoming increasingly clear that plants have sensory systems and response effectors that are functionally similar in important ways to those that we find in animals, even "higher animals" such as fish, birds and mammals. So, one should not slip

DOI: 10.4324/9781003393375-8

into the once popular notion that plants are some kind of reflex machine. Yet, it is possible to go too far in the opposite direction. Some treat plants as if they were sessile versions of higher animals. But this neglects some important differences. In this chapter, I am concerned with one important difference. I argue that certain animals possess a sensory function that sets them apart from plants in a significant way. As we will see, this function is at the root of one of the most significant elements of the human self-conception.

5.2 Othering Impressions: Sensing vs. Perceiving

I have two chief aims here. The first is to show that plants and other non-animals *sense* in much the same way as animals—that is, their sensory systems have many of the same functions. My second aim is to display a difference. I will argue that plants are not capable of *perceiving*—only animals have this additional capacity.

Here is what I mean. I will say that:

A subject S *perceives* an entity X when S senses X as being distinct from S.[2]

Here are some examples. I (S) *perceive* when I feel *a heavy spherical object* (X) in my hand or hear *a loud squealing sound* (X') outside the window or catch sight of *a shady place to rest* (X") just down the road. (By contrast, when I feel hungry, I sense something, but not something distinct from me.)

In each of these sensory states, I am presented with an entity that appears to be distinct from me. For first, each of these objects appears to be located in a common space beyond the boundaries of my body, albeit in a common space in which my body also appears located. This means that if the above perceptions are concurrent, they appear to be located both relative to me and relative to one another in this space. (I am asserting this to be a sufficient, but not a necessary, for distinctness.)

Moreover, these objects may appear to be things my body can affect in certain ways. For example, I feel that I am supporting the heavy spherical object in my hand and preventing it from dropping out of my grasp. These appearances are of the objects in a sensorimotor framework, a "body schema" (Gallagher 1986, 2001) that enables me to handle and manipulate them.

Audition and vision also present me with othering impressions. That is, the things that I hear or see appear to be at a distance from me, located in a common space, and it seems that I can approach or move away from them.

These sensory states have the force of their objects being distinct from me. I cannot drop or approach something that is a part of my body. Things located in positions other than mine in a common space cannot be identical with me. Call them *othering impressions*.

To sum this up, I define *perception* as sensing that presents an othering impression of some entity. Mere sensing—sensing that is not perceiving— is not so marked. My thesis is that plants sense but do not have othering impressions. They do not perceive and have no use for perception. Only animals perceive.

5.3 A Little History: Errors along the Way

Before I get into my main line of argument, let me recount some history. Botany is a neglected subject in philosophy. Very few historical texts discuss plants, but when they do, they mostly pose the question of plant sensing in terms that suggest *perception*. As we shall see, modern scientific investigation does not conflate the notions in this manner. This corrects a bias that defines sensing in terms of othering and points to a conception of sensing that is more inclusive of plants. Just a little schematic history will help us understand the conceptual innovation that I am suggesting.

The dominant European tradition derives from Aristotle, who held that plants were alive but could not sense.[3] Aristotle attributes to plants a "nutritive" or "vegetative" soul, but not a "sensitive" soul.

> Each animal insofar as it is an animal has the capacity to sense, for it is by this that we distinguish between what is and what is not an animal.
>
> (*de Sensu* 1, 436b10-12)

In short, plants are by their very nature—i.e., by what it is to be a plant— unable to sense.

How did Aristotle arrive at this view? He seems to think that creatures that sense are subject to (a) hunger and thirst, which are forms of desire (*de Anima* II 3, 414a32-b16) as well as (b) sleep. Sleep is not relevant here, so let's put it aside. Desire impels creatures to move, Aristotle says; there is thus no role for it in a sessile organism. This is why plants lack the power to sense, according to him. Notice how this is framed in terms that are better expressed by our term "perceive." Plants are not capable of desire; desire is directed to objects at a distance; hence they do not sense anything that is at a distance. But since this argument allows that they might sense their own states, the properly drawn conclusion is that they don't *perceive*, because they don't sense anything outside.

In India, the question of plants was more widely discussed, and opinion was more divided. Many Indian philosophers realized that plants are alive. According to the Upanishads (followed by the Jains in this respect) they are "sprout-born," though many Buddhists said that they were not "breathing beings," which caused them to equivocate about their animate nature (Findly 2002). Quite commonly, they are designated as *ekindriya*, or "possessors of a single faculty," that of touch—presumably, on the grounds that they are sensitive to conditions that (in humans) are sensed by tactile contact. According to Findly, there is a tradition originating in the *Mahabharata* that ascribes "all five sense faculties to plants"—a tradition also taken up by the Jain philosopher, Jinabhadra, who says (as Piotr Balcerowicz puts it in private correspondence) that plants "experience (or react to) colour, sound, smell and taste through special capacities."[4]

One point of interest to note right away is that both traditions identify sensing with the external sense modalities by which animals perceive—primarily, the traditional five. For them, the question "Do plants sense?" is just the question, "Do plants touch, hear, see, smell or taste?" And we have three answers to this question.

(1) Aristotle says no, they do not touch or hear or see, etc.
(2) Most Jains and some early Buddhists say that they exploit touch but no other modality; plants are *ekindriya*.
(3) Jinabhadra says they have all five senses.

However, posing the question in this way—i.e., in terms of the five senses that animals possess—is already an error, as I'll argue in a moment.

But before I get to this, let me note a further, equally damaging error. Since verbs of perception normally take a direct object, the question whether plants touch, hear see, etc. becomes "*What* (if anything) do plants touch, hear, see, etc?" By this very natural-seeming slide, this transforms the inquiry into one about external perception. For Aristotle, who (rightly) links external perception and motility, plants cannot see, hear, touch, etc., because they have no use for awareness of edible things at a distance from themselves, things that they desire to move towards or away from, and so on—they don't sense these external things, and so they don't sense at all. For a philosopher like Jinabhadra, however, the question is one of *sensitivity*—for him, I assume, the question "Can a plant see?" reduces to the question whether plants are sensitive to things that animals like ourselves see. Since they are—they are sensitive to light and dark, for example—a philosopher like Jinabhadra would be disposed to conclude that the question has a positive answer. They see, and it follows that they see something. Minimally, they see light and darkness.[5]

Here's the thing, though. An organism might be sensitive to some stimulus without perceiving that stimulus. For example, it might be sensitive to darkness, but this sensitivity might express itself simply by making it sleepy when it gets dark. In other words, it might sense darkness and thus become sleepy without sensing *that* it is dark. In both Europe and India, the transitivity of perceptual verbs obscured this possibility.

The fallacious inference from sensitivity to X to perceiving X is still with us. For example, Paco Calvo and Karl Friston (2017) powerfully argue that, as they put it: "Plants respond in a fast, and yet coordinated manner, to environmental contingencies" (ibid., 1). This is surely right—it is agreed territory. Calvo and Friston argue that predictive processing is involved here. In other words, the system responds to a proximal stimulus in a manner befitting a distal stimulus anticipated in the near future. They write: "plant perception entails predictive hypotheses *as to what is out there*: it could be a herbivore representing bad news, it could be a stream of water, or what may" (4, emphasis added). What they have actually established is this: at time *t*, a plant may initiate a response that is appropriate to water (or some other distal stimulus) at t + ∂. Let us grant that this might be true. It still does not entail that the plant locates the water in a body schema, or common space, or that in some other way others it. Sensing something is not necessarily to have an othering perception of it.

Aristotle and Jinabhadra have opposed views on the subject of plant sensing—one thinks that lack any sense faculty, including the most basic one of touch, while the other thinks that they possess them all. And so, at first glance, it seems that Aristotle has a demanding criterion of sensing—that it must be associated with desire, for example—and that Jinabhadra has a permissive criterion—one that demands only sensitivity to the stimuli characteristic of a particular modality. Their positions are appropriate to their criteria, but the fact remains that neither poses the question properly. The question should *not* be: Do plants possess *our* five external modalities? This encourages an inappropriate zoocentrism. When considering plant sensing, the question should rather be:

> Do plants have dedicated systems that are (a) sensitive to circumstances that require differential response for the maintenance of life, and (b) capable of initiating response to these circumstances as they occur?

This question is more accommodating. It allows for specialized sensory mechanisms that have nothing to do with the traditional five.[6] And it does not require othering representations of sensory objects.

5.4 Scientific Investigations

The modern conception of sensing seems to emerge slowly from scientific investigations of plant behaviour from the late eighteenth century onwards. The investigators of this period were looking at how plants adapt to change. This was widely thought to be mediated by sensory organs.

In 1779, the Genevan philosopher of nature, Charles Bonnet reported[7] that if you lay a potted plant on its side, its stem bends and reorients to the vertical as it grows. Bonnet believed in a chain of being that ascends from "subtle matter" at the bottom to humans at the top. According to him, powers of sensing emerged in plants, which he placed nearly half-way up his chain. He took the gravitropism of his potted plants to be evidence of this. We now know more about the sensory capacity that mediates this behaviour. Plants have cells known as statocysts, which contain organelles known as statoliths. When these cells are tilted off the vertical, the stato-liths "sediment." This change induces the statocyst to emit an ionic signal. This initiates the turn of the stem. Thus, the statoliths function as receptors and the statocysts as transducers that emit sensory signals for gravitropism.

Following investigators such as Bonnet, sensing came to be seen as a mediator of variable response with no assumption made about othering impressions. The new attitude is evident in the work of Charles Darwin and two of his sons who, building on the work of a number of botanists,[8] conducted extensive experiments on plants. (Figure 5.1 is a trace of plant movement by the Darwins.) These culminated in a book entitled *The Power of Movement in Plants* (1880) with the stated aim of giving a uni-fied account of "several large classes of movement, common to almost all plants."

A plant's growing parts spread in ways that benefit the plant. Its roots must overcome the resistance of the earth; its stem grows vertically upwards; its tendrils need to find support; its leafy branches seek light. The Darwins realized that these movements were actively managed.

> When light strikes one side of a plant, or light changes into darkness, or when gravitation acts on a displaced part, the plant is enabled in some unknown manner to increase the always varying turgescence of the cells on one side; so that the ordinary circumnutating movement is modified, and the part bends either to or from the exciting cause.
>
> (1880, 547–8)

The primary movement of the growing parts that Darwin studies is "cir-cumnutation," the more or less random circling of roots, tendrils, etc. "to all points of the compass" (ibid., 1)—"every growing part of every plant is continually circumnutating, though often on a small scale" (ibid., 3). Circumnutation is an underlying random substrate of directed move-ment—random in the sense that it is not *toward* or *away from* any source of benefit. Additional motions are superimposed on circumnutation in accordance with need; a root's circumnutation is biased in the direction

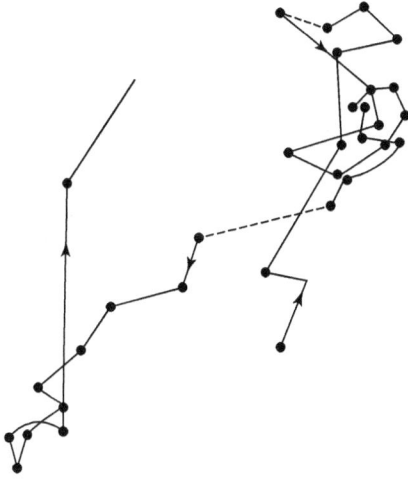

Brassica oleracea: circumnutation of radicle, traced on horizontal glass, from 9 A.M. Jan. 31st to 9 P.M. Feb. 2nd. Movement of bead at end of filament magnified about 40 times.

Figure 5.1 The movement of a cabbage radicle, one of many such tracings in Darwin (1880, 11). Darwin established that the tip of a radicle received "impressions from the sense-organs" and directed the movement of the shoot.

of gravity, for example, and a shoot's is biased in the opposite direction. "There is always movement in progress, and its amplitude, or direction, or both, have only to be modified for the good of the plant in relation with internal or external stimuli" (ibid., 4).

Darwin realized that movements directed by "internal or external stimuli" implied sensing those stimuli. He attributed sensory function to the tip of a plant's radicle. Amputate this, and the root's environmental responsivity is eliminated, and it behaves in an inflexible way.

It is hardly an exaggeration to say that the tip of the radicle [...] having the power of directing the movements of the adjoining parts, acts like the brain of one of the lower animals; the brain being seated within the anterior end of the body, receiving impressions from the sense-organs, and directing the several movements.

(ibid., 573)

This is Darwin's ground for attributing sensory function to plants.

The message we should take from Darwin is that all living things respond in a functionally appropriate way to occurrences and circumstances that affect their well-being. In order to be able to do this, organisms generally have to transition to an "effector state" that initiates action appropriate to the circumstances they encounter. To achieve the right match between effector states and prevailing circumstances, organisms need to possess receptors that are receptive to the chemical, mechanical and electromagnetic influences of their environment and organs that transduce these influences into signals that initiate their response. In short, they need sense receptors, transducers and sensory systems.

5.5 Plant Behaviour Is Complex but Does Not Employ Othering Representation

Tye says that plants are "tropistic organisms." Is Bonnet's phenomenon a "tropism?" Yes, because a tropism in plants is defined as "directional growth in response to a directional stimulus" (Gilroy 2008, R275). Gravity is a directional stimulus, and plants grow directionally in response. But one should not assume (as Tye apparently does) that tropisms are simple, genetically determined reflexes. In fact, they are quite complex. Gravitropism in particular is not a simple bend to the vertical. The upward bend of a shoot is achieved by adding extra growth at the lower side. (Plants bend towards one direction by adding auxin, the growth hormone to the opposite side of the stem.) But the shape of the shoot is maintained throughout; the growth on the lower side of the bend does not carry through in the form of a lump or thickening. Accordingly, there is a "decurving" process that ensures that the shoot *above* the bend has normal dimensions. In addition to decurving, there is also phototropism and other natural tendencies. Each of these processes requires input from dedicated sensors (Bastien, Douady, and Moulia, 2015; Moulton, Olivetti, and Goriely 2020). As a consequence, plant behaviour integrates the output of multiple sensors. As one recent paper on gravitropism in wheat coleoptiles notes:

> [C]oleoptiles respond not only to sums but also to differences between stimuli over different timescales, constituting evidence that plants can compare stimuli—crucial for search and regulation processes.
> (Rivière and Meroz 2023, 1)

Coming to another sensory complexity, plants appear to learn. Pea tendrils coil if they sense contact in the light but stay uncoiled if they are rubbed while in the dark (see Engelberth 2003 for a review). In a classic study, M. J. Jaffe (1977) showed that the effects of rubbing are "stored"; tendrils that were rubbed in the dark immediately coiled when exposed to light up to two hours later. Gagliano et al. (2014) found that when *M. pudica* (see note 8) is subjected to continuous and identical disturbances, such as drops of water, it modulated its initial response, which was to close it leaves. Gagliano and her co-authors suggest that this modification of behaviour "exhibits clear habituation, suggesting some elementary form of learning" (ibid., 63; see also Affifi 2013).

There is no doubt, then, that plants are able to detect changing environmental conditions, and that their responses "greatly transcend the phenomenon of cellular irritability" (Charles Binet, quoted by

Trewavas 2014). Plant behaviour is not, in other words, a simple passive reaction to events on their surfaces (as Aristotle and his successors believed); it requires at least the classification and comparison of these events and the activation of internal processes.

Though plant sensing is complex in these ways, I want to argue that it does not require othering representation of the outside environment. Even if the statoliths are presentations of the direction of gravity, and even if gravity is outside and distinct from the plant, plants do not have othering representations of gravity or its direction. Their growth behaviour does not, in other words, require gravity to be represented as something outside and distinct from the plant. This, I suggest, is universally true of plant sensing. The sensory regulation of plant behaviour does not depend on making a distinction between self and other. It does not require othering representations.[9]

Now, this might seem counter-intuitive. Many years ago, I argued (Matthen 1988) that:

A perceptual state is a presentation of F if and only if its function is to detect F.

This "teleosemantic" approach to representational content implies that plant sensors represent external circumstances—if a sensor drives response to an external circumstance, it leads us to say that it represents that circumstance. The function of statocysts is to drive response to gravity. So, should we not say that statocysts represent *gravity*, which happens to be a force external to the plant? Is this an othering impression? Not necessarily. For representing an external circumstance does not entail representing it *as* external. The plant changes state when certain external circumstances are detected, but there is nothing in its state that makes it respond as to an external thing. As complex as it is, then, the gravitropism described above does not imply that gravity is represented as a force outside the organism.

5.6 Perception and Othering in Higher Animals

Higher animals *perceive*. That is, their sensory processes yield othering representations of entities located by means of "cognitive maps" of the space outside the organism itself. When I take my seat in a crowded subway car, carefully maintaining a polite distance from the person already sitting there, I am acting in sensory awareness of this other person and where she is, relative to me and also relative to the other things and people I see.

There are two features of othering awareness that are important here.

First, I am aware of the things around me as items that I could bump into or otherwise interact with. That is, I possess a sensorimotor awareness of these things, relative to my limbs and their movements, including a space-like framework that enables my limbs to effect action on nearby things. Shaun Gallagher (1986, 2001) called this framework the "body schema"—a sensorimotor system that "constantly regulates posture and movement [...] without the necessity of perceptual monitoring" (2001, 149).[10]

And second the things I externally perceive appear to me as possessing metric location relative to one another and to me. This implies a "cognitive map"[11] or distance- and direction-based location scheme for everything externally perceived (including my own bodies).

The bodily scheme and cognitive map are the foundations of othering impressions—these things are presented as distinct from me in consequence of the above spatial features of their appearance.

Why *spatial* representation? The crucial point here is that higher animals are freely motile: they move through their environment towards or away from points of their choosing. They keep track of the things they have to navigate to, around, and away from. Plants, by contrast, are sessile: they merely "grow through their environment" (Gilroy 2008, R275). In higher animals, such as birds and mammals, motility brings environmental challenges that require locational awareness and navigational skills.[12] This is why these animals perceive—why they have sensory representations of things as located in the space outside themselves.

The contrast that I am making here provokes a sceptical question. Concede that plants act on the basis of the condition of their own sensors. After all: what other source of information do they have? But, says the sceptic, isn't exactly the same true of animals? We experience visual, auditory, tactual, olfactory, gustatory impressions. These impressions are the products of our senses. Since we have no other source of information, we must be acting on the basis of these alone. (This, in fact, is the point of view of Thomas Reid, to whose distinction between sensation and perception I am indebted—see note 1.) How are we different from plants? Whence do othering impressions arise?

The answer to this question is, perhaps, surprising. Birds and mammals (and possibly have *innate* spatial frameworks—they have inner representational organs that impart spatial form to sensory output. They do not place things in external space by somehow inferring spatial relations from non-spatial sense impressions. Rather, their sensory organs are genetically constructed to place all external objects in a spatial matrix with themselves at the origin or universal reference point. With respect to sensorimotor action, this matrix is Gallagher's (1986, 2001) "body schema." With respect to navigation, mammals and birds (and perhaps some fish[13]) array external perceptual sources in the hippocampus. This organ serves as a map of the external sources relative to the organism itself. Both kinds of representations are othering because they locate things at measured distances away from the organism. Plants have no facility that generates othering representations. They operate just on the state of their sensors without sorting out the self and other.

5.7 What Is Sensing?

In light of the discussion so far, I offer the following as a definition of sensing *without* any assumption of othering representation.

Sensing is a capacity S of an entity E to:

    a be in a state mediated by distinct receptive and transducing organs that

    b covaries with variable circumstances, where

    c each such state has the biological function of triggering E's normal response to the circumstance to which it corresponds.

The definition is apt for:

(1) the traditional cases, where the responses are epistemic. For example, colour vision is the capacity of an organism to be in different states for the different colours that it can discriminate by means of colour sensors, where each of these states has the biological function of triggering different belief tendencies about the colour of seen objects.

The definition also fits:

(2) sensory capacities that serve bodily regulation. For example, thermal sensors are in different states when exposed to different temperatures and serve to trigger thermoregulatory actions (such as sweating or shivering) appropriate to those temperatures.

And it accommodates:

(3) sensory systems (such as appetite) that generate motivational states (such as hunger) that impel an animal to act in a manner that is appropriate to the circumstance detected.[14]

5.8 Chance: Growth that Goes against Gradients

Now, let's consider behaviours that suggest spatial representations in plants and why these behaviours do not in fact require spatial representation.

Plants send roots towards areas of greater moisture and nutrition content. If a root is growing towards such an area and encounters an obstacle, it grows around the obstacle and resumes its growth towards its initial destination. How do they determine where these better areas are? It might seem that in order to do this it must explore its surroundings with the aim of finding the resources it needs. And this would imply that it represents the spatial region around it. I will argue that this is not so. There are two forms of behaviour that achieve the said result while only requiring a sampling of proximal values within, or at the surface of, an organism's body. These do not require a representation either of external space or of the locations of distal objects in external space. Plants act simply on the basis of the state of their own sensors.

First, I'll give an account of gradient-based behaviour, or tropisms. Think of a *field* as a spatial distribution of some physical variable: for instance, electromagnetic strength, gravitational force, velocity and volume of water flow, chemical concentration and so on. What do you do when you are trying to track down the source of a foul odour in your kitchen? You sniff and move, following your nose to higher odour concentrations. In effect, you sample and compare odour field strength at various points and follow increasing or decreasing strength depending on your purposes. The object that interests you is located where the field strength is greatest. You rely on *field gradients* in order to get to or away from it.

How, in general, does an organism detect a field gradient, given that its sensors can only detect the strength of the field at a given point? There are two possible ways. The first is the one just mentioned: it samples field strength at successive locations and compares these readings over time. The second is that the strength of its gradient-directional orientation is determined by the strength of its field vector reading. For example, it might be more strongly pulled in the direction of the vector when the field is stronger. Thus, where the field is weak, it might waver with respect to that direction, but as the field-strength increases it might fix on the gradient with less and less wavering. This second method requires only one reading at a time—no comparisons over time. I suggest that the second would be the one followed by plants, there being no evidence of them being able to make comparisons over measured time.

Gradient-based change has been extensively discussed in the literature on evolution by natural selection, raising issues that are relevant to us now. Natural selection is gradient based on a metric constructed in the following way. Graphically array all possible genotypes available to a population of organisms so that similar genotypes are close to one another along

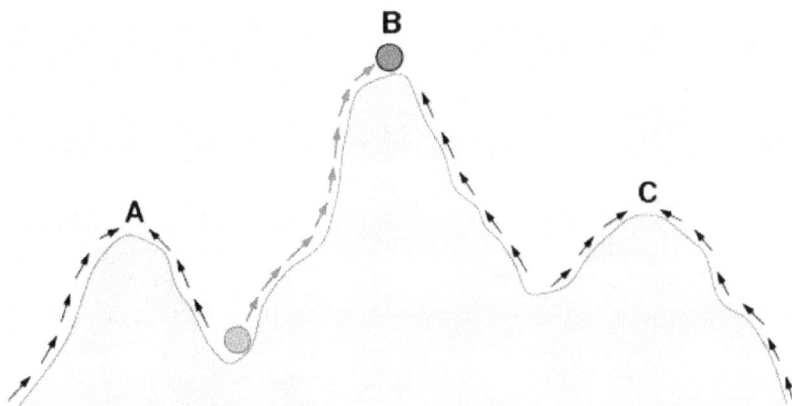

Figure 5.2 A diagram of fitness values (y-axis) that attach to phenotype variation (x-axis) in a population. (Source: https://en.wikipedia.org/wiki/File: Fitness-landscape-cartoon.png)

the independent axes of the graph and dissimilar ones far apart.[15] Each genotype has a fitness-value, and this is represented in the graph as the "height" of that genotype along the dependent axis. Figure 5.2 is a simplified example of such a graphical representation of fitness: it assumes only one dimension of similarity. The "landscape" shown in Fig. 5.2 is in effect a field-map that represents fitness as a function of a single genotypic variable. Populations will (with high probability) transition from states of lower fitness to adjacent states of higher fitness but will *not* (again with high probability) transition from states of high fitness to adjacent states of lower fitness. This is a consequence of the Principle of Natural Selection. Looked at in this way, natural selection is a tropism—the directional movement of a population in response to a directional stimulus.

Now consider the population marked by the circle near the trough between A and B in Fig. 5.2. At the trough, this population is at a local minimum, meaning that its fitness is lower than that of any similarity-adjacent point. Any mutation, or change in its genotype, would result in an increase of fitness. Now, remembering that this landscape varies in only one respect, and assuming for the sake of simplicity that the slope of the fitness curve is equal in both directions, this population is as likely to move towards B as it is towards A. Once it starts up either slope, it has only one way to go to increase its fitness, because any change in the opposite direction would then be a decrease in fitness. So, if we were to assume that populations *always* ascend a fitness gradient, some populations in the trough of a curve like that of Fig. 5.2 will miss the global maximum. On our assumptions, each such population is equally likely to ascend only to a local maximum, such as A, even though this is not the highest fitness point in the entire landscape. And an ascent to A is irreversible on these assumptions. There is no way it can get from A to B without first reducing its fitness, and natural selection makes it highly unlikely that any such reduction could occur.

This yields a general point that was made by Jon Elster (1979), which I recount in David Gauthier's (1983) felicitous summary:

> Although natural selection yields local maxima, it does not in general yield global maxima [...] it can not wait; it can not refuse a favourable mutation so that it may later be able to accept a more favourable one. And it cannot accept an unfavourable mutation [that takes it downhill from A] so that it may later accept an even more favourable one [that takes it upward to B from the trough between A and B].
>
> (ibid., 134)

According to Elster, the achievement of non-local maxima requires a representation of the distant terrain. If a population at A could represent B as a distant non-local maximum, then it could accept an unfavourable mutation in order to get to B. But this is, of course, what a population under natural selection cannot do. It does not represent the whole fitness landscape; it merely moves along fitness gradients (Fig. 5.2).

One might reach a similar conclusion about the gradient-based growth of a plant's root system. Reinterpret Fig. 5.2, now, as the nutrient concentration height of adjacent locations. The ball in Fig. 5.2 is now the position of a root tip following a nutrient gradient. Suppose that the root found its way to A, a local nutrient maximum. Then its path to B would be blocked. This happens more frequently than you might think. Suppose a root is following an ascending nutrient gradient when it comes upon an impenetrable obstacle—a granite rock, say. Its best course is to grow around the rock and pick up the interrupted gradient on the other side. But by Elster's argument as presented above, it cannot do this. It cannot accept a reduction of nutrient concentration so that it may later accept a more favourable increase. Any change of direction would be along a negative gradient. In effect, the place where it meets the rock is a local maximum—every step taken from there represents a reduction of nutrient concentration. Consequently, the gradient following root is simply stuck there. The problem is, of course, that root systems *don't* get stuck when they meet obstacles. They *do* go around them. So how does this happen?

Situations of this type were debated by R. A. Fisher and Sewall Wright. Both thought that it was indeed possible to get from A to B in Fig. 5.2. Both their views are relevant. Fisher's idea was that it was easy to underestimate the dimensionality of the fitness landscape. Figure 5.2 records fitness values for one dimension of genotype variation. But what if there were another dimension. Then, there could be an ascending path from A to B that lies "behind" the plane diagrammed above. Call this the "added dimension" solution.

Wright, for his part, pointed out that genetic drift could solve the problem. Evolving populations engage in a fitness-weighted "random walk"; they do not evolve monotonically to the fittest alternative, but rather fluctuate back and forth in the overall direction of that alternative. If selection were deterministic, a population could never descend from A in order to get to the global maximum, B. However, chance occurrences could make it drift to the trough between A and B and then ascend to B. Call this the "random walk" solution.

Both these solutions are relevant to the gradient-based directionality of growth that we have been discussing. Consider Fisher's added dimension solution first. We assumed that the root would follow a nutrition gradient, but as we have seen, there are other relevant gradients for it to follow. The root that is stuck at the granite obstacle might have no ascending nutritional gradient to follow, but at that point it might follow an ascending wetness, gravitational, or (as it actually happens) thigmotropic or pressure gradient.[16] In fact, it might all along have been following an integrated gradient that does not peak at the rock obstacle—we noted earlier that plants integrate their sensory inputs. Sensory integration creates gradients in multiple dimensions.

Now consider Wright's random walk solution. A root that has nowhere to grow but downwards on a nutrition gradient may grow in random directions. This too is consistent with plant behaviour. Just as an evolving population moves about the fitness landscape randomly with an overall direction superimposed by selection, so also a plant's roots and tendrils constantly "circumnutate." As we saw, Darwin thought that movements along field gradients were superimposed on circumnutation. Circumnutation can carry the root anywhere, but sensed gradients bias its growth in the direction of field gradient vectors. The stronger the field, the greater the bias towards the field-gradient. Like a tracking dog its growth would waver back and forth at first but become straighter and straighter as the field became stronger. This could carry a root tip down a field gradient to a point where it has a clear path to a global nutrition maximum.

To summarize, then, a root that encounters an obstacle along its ascending gradient could either grow along another gradient or grow randomly until it found a point from which it could resume its ascending path. Elster supposed that natural selection, operant conditioning and supply-and-demand market forces are governed entirely by field gradients and are thus doomed to get stuck in local maxima. (He thought, by the way, that this is an argument in favour of state planning.) But he was wrong: gradients of higher dimensionality and random walks disturb a system's stable equilibrium at local maxima, clearing their path to non-local maxima.

The conclusion that we can draw about plant behaviour is that internal controls are, first of all, multidimensional. And in addition to all of the force fields that constitute these dimensions, there is an underlying random motion on which the force fields are superimposed. Plant growth integrates all of these.

5.9 Object-Targeted Behaviour

Earlier, I said that certain organisms have no interest in *objects*. The root system of a plant doesn't care where the nutrition in the soil comes from—e.g., a dead animal that is decomposing in the soil or a drip of moisture that is dissolving minerals in the soil. It simply maximizes (or rather optimizes) the nutrition available to it. However, there are plant behaviours that seem to target objects. Do these behaviours require a representation of these objects?

Consider the behaviour of carnivorous plants such as the Venus flytrap. It is easy to think that they *are* concerned with the insects that alight on their leaves; their behaviour is as much directed to eating their prey, as mine might be to the buffet table at the other end of the room. But the flytrap doesn't have to represent the insect or its location in order to attain its goal. Here's a description of how it operates:

> the whole central zone which is covered with digestive glands. Among these are bristles known as trigger hairs, usually three to a lobe. When an insect crawls along the ventral surfaces and bumps into the three small trigger hairs, the trap snaps shut. Two touches of a trigger hair activate the trap which snaps in a fraction of a second at room temperature.
>
> (Pavlović, Demko, and Hudák 2010, 37)

Here's how this could work without a representation or measurement of outside objects, or of space and time. Let us say that the flower is in a "resting state" R. At this point, trigger hair H is touched. H responds by causing the flower to enter into an "alert state" A. When the flower is in state A and a trigger hair H is disturbed, the plant's leaves snap shut. Since the alert state decays rapidly, the trap shuts only when trigger hairs are disturbed twice in a short time period.

The above can be represented as a machine program for trigger hairs:

If the flower is in state R, and trigger hair H is disturbed, then switch the flower to state A.

If the flower is in state A and trigger hair H is disturbed, then activate the snap-shut routine.

If the flower is in state A for more than t (t $\cong$ 20s), then revert to R.

Each line of the program involves a conditional action. We may suppose that these are realized by chemical means; for instance, the switch-to-A might be achieved by the rapid propagation of some chemical, and the reversion to R by the decay of this chemical. (The snap-shut routine is more complex.) In this way, sessile organisms can react to a spatiotemporal pattern occurring on their bodily surfaces (or interiors) without needing to represent the outside causes explicitly. Their internal states mediate behaviour that is sensitive to spatiotemporal patterns without representing those patterns. There is no need for the Venus flytrap to sense the fly as distinct from other objects in the vicinity; there is no need for it to represent where the fly is.

5.10 Targeted Motility and Its Demands

As complex as detection and choice functions are in sessile organisms, organisms that move themselves from one place to another are faced with problems that demand a different kind of sensory function. Organisms that are motile in this way need and are equipped to search for food not just in contiguous soil and air but also in regions that are some distance removed from themselves. I walk across the crowded reception room to greet the friend I haven't run into for a long time. In doing this, I not only make choices based on what I perceive distally but also plot a course through space to the person I thus single out. I don't follow a gradient in order to do this; I plot a course to an object identified at a distance and correct my course as I go. Behaviour that targets distant objects demands a different kind of perceptual representation of space.

Let us now consider a particular form of motility, which is likely what Aristotle had in mind when he said that the soul "pursues and flees" sensed objects that are pleasant or aversive. *Targeted* motility, or *t*-motility, is self-actuated, self-propelled motion *toward (or away from) a distant object or location* through terrain that might require an indirect path around obstacles or course-correction as new circumstances arise.

I'll give three kinds of evidence that shows that animals are *t*-motile.

Start with the phenomenological. Organisms that are *t*-motile are equipped to search for food not just in contiguous soil and air but also in regions that are some distance removed from themselves. Behaviour that targets distant objects demands a perceptual representation of space. Let's suppose that you start from A and set a course to B, which you can initially see. The straight path to B is blocked and so you have to plot an indirect trajectory. Moreover, there are objects along the way that obscure your view of B when you are up close to them (see Figure 2.2.3) (Fig. 5.3).

t-motile organisms keep track of B while detouring around the objects in the way. At many points, B is out of sight behind the obstacle, but they are able to track their own position relative to B despite this. They do this by updating their position on an internal map on which the position of B is marked. Input to the update is perceptual movement information. This information is in a form that allows it to update the map. They cannot achieve this by sampling the strength of some field that maxes out at B.

The above was phenomenological evidence. It seems, subjectively, that we are capable of *t*-motile action. Now here are two performance-based criteria.

The first such criterion was devised by E. C. Tolman (1948) and colleagues (Tolman, Ritchie, and Kalish 1946). It is meant to show that animals represent their surroundings in the form of a "cognitive map," a mental representation that enables them to store and recall geometrical features of recalled environments.[17] As Michael Rescorla (2018) reports, scientists defend the existence of these maps by showing that animals are able to "take novel detours and shortcuts" (ibid., 35). "A recurring experimental paradigm [...] is to displace the animal to an unfamiliar point within a familiar environment. In many cases, the animal travels directly from the release point to the goal" (ibid.). For example, Tolman, Ritchie and Kalish (1946) trained rats to find food in a maze like that in Fig. 5.4.

Having done so, they then removed the maze, replacing it with an apparatus of radiating chutes with the starting point and food reward unmoved (Fig. 5.5).

The result was that the rats were to run straight to H or at least come close. This shows that they could, in Rescorla's words, take novel shortcuts.

C. R. Gallistel (1990) relates a similar performance in a human child. He recounts how Barbara Landau and Lila Gleitman (1985) brought a congenitally blind child of 31 months into their lab and guided her from a point of

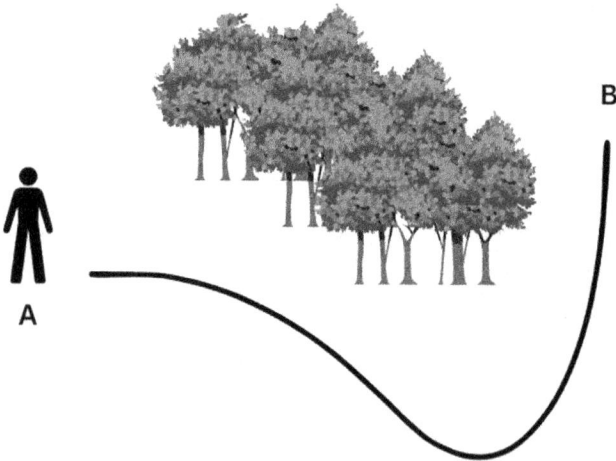

Figure 5.3 Finding your way around an obstacle.

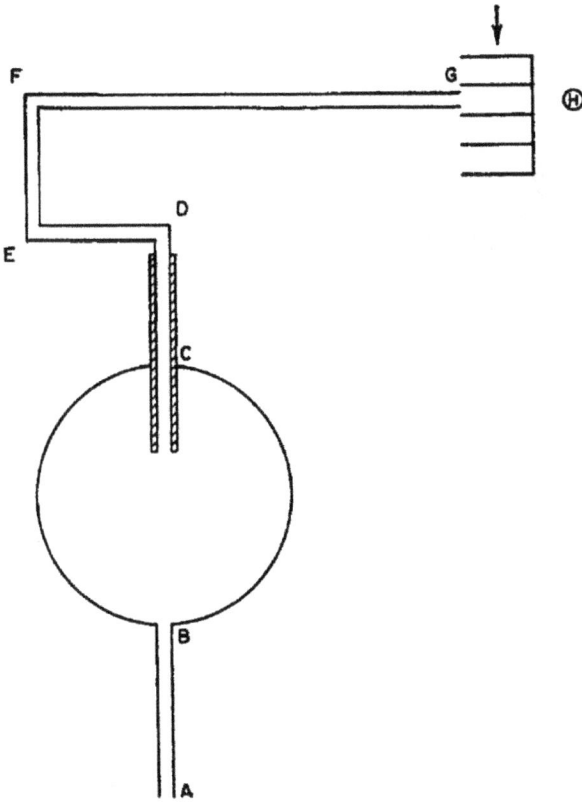

Figure 5.4 Maze used to train rats to find food located at point H.

origin (where her mother was seated) to various objects in the lab—a table, a basket and so on—always bringing her back to Mother after visiting each object. Once the child had become familiar with the paths from Mother to each object, they guided her to one of the objects, the table, and asked her to find her way to another object, the basket. She was able to trace a more or less straight line between these objects, not having to travel via Mother. Gallistel remarks: "In holding its course, the child was presumably entirely dependent on dead reckoning, calculating changes in her position from vestibular, kinaesthetic, and efference copy signals" (1990, 101). The child did not find the basket by muscle memory; she had no muscle memory of the table-to-basket path. She also did not use a field created by some signal emanating from the basket; in particular, she could not use the fact that the destination object loomed as she approached it. Rather, her spatial sense was able to construct an internal map of the objects from the paths traversed earlier, which she then used to trace a direct path from one object to another

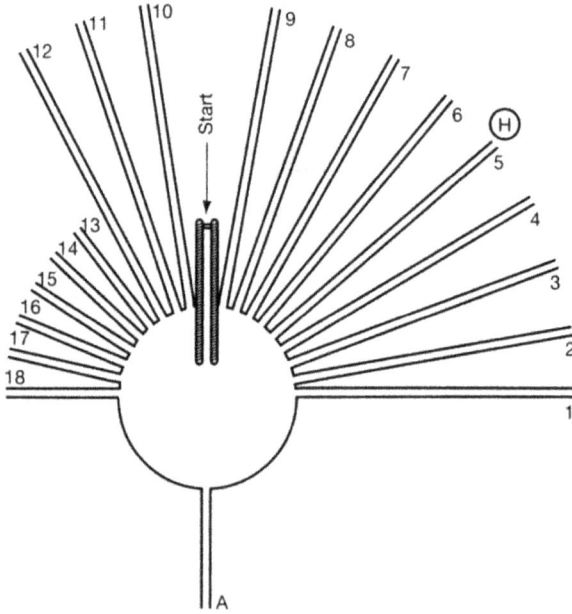

Figure 5.5 Radiating chutes that replace the maze shown in Fig. 5.4. Note the start-point on the circular platform and the end-point, H, as in Fig. 5.4.

by-passing the earlier circuitous routes via Mother. She used motion signals from her vestibular, kinaesthetic and motor systems to encode her own movement vectors to distinct located objects; she had no access to a field strength gradient for the purpose.

There are many other phenomena of this sort. Foraging animals—ants, bats, honey-bees—take a random exploratory path from their home base in search of food. When they have found food, they are able to return to their home by a direct path.

Finally, there is direct evidence from the brain of cognitive maps contained therein.

John O'Keefe and Lynn Nadel (1978) and co-workers showed that the hippocampus contains an array of so-called "place cells," each of which is associated with a specific place in the environment. Workers in later years, many in O'Keefe's research groups at University College, London, slowly discovered that these place cells are supported by a variety of location-indicating cells such as grid cells (which fire when the animal is at an intersection of lines that make up a grid that covers the environment), border cells (corresponding to walls and other limits to the animal's environment) and head-direction cells (which mark the animal's heading).[18] Together, the array of place-cells and border-cells thus constructed constitutes an allocentric map of a remembered place, with head-direction cells providing a translation from egocentric to allocentric coordinates.

The hippocampus is an organ that structures sensory input from the external senses in a matrix that corresponds to external space, thus enabling us to perceive objects as possessing allocentric location in a single framework. Recall the sceptic who questions how it is possible for an organism to possess information that goes beyond what is given by sensory receptors. The answer is that the hippocampus imposes spatial structure on this information. This is not an arbitrary imposition. The spatial structure works to provide the organism with the phenomenological structure of t-motility which, in addition, meets the performance requirements sketched above. As O'Keefe and Nadel (1978) argue, the hippocampus is an organ that imposes a Kantian a priori onto the sensory substrate of the eye, ear, and so on.

5.11 Conclusion

In perception, animals make contact with a world arrayed in three spatial dimensions. They do so in two ways. First, all motile animals have a body schema for sensorimotor interaction. Second, higher animals such as teleost fish, birds and mammals have cognitive maps for the allocentric location of perceived entities. Spatial sensory states of this sort are othering—entities appear distinct from the perceiving subject in virtue of perception.

Plants do not perceive; they simply regulate their own bodies in accordance with programs based on the input of their sensors. We should not underestimate the claims of such self-regulating sensing. Plants are not passive lumps of organic matter. But we should not overestimate it either. Plant sensing is not the same as external sensing, consciousness or agency. It is entirely self-regarding.

When humans are conscious, they are self-conscious. When I think, I have an idea of myself as the agent of thinking. Such a mode of thought implies a distinction, an idea of *non*-self. When I think of the outside world, I think of it as real inasmuch as I think of it as something outside myself. These notions of self, of self and other, and of independently existing reality are all founded on sense impressions that present the targets of action as outside the agent's body. These impressions are the "othering" impressions that I have discussed in this book. These impressions are what set perception aside from mere sensation. Only animals are capable of perception. If plants are conscious, then they are solipsists.

Acknowledgements

I am very grateful to Peter Schulte and an anonymous referee for extremely helpful and constructive comments.

Notes

1 I would like to say explicitly what may not be obvious because of this criticism. I believe Tye's work on perception is generally admirable.
2 This definition is inspired by, but a little different from, Thomas Reid's (1785/2002) classic conception. According to Reid, a *sensation* is a sense impression I that has no object distinct from I itself (*EIP*, I 1 12). A perception, for him, is a mental state P formed on the basis of sensations that has an object distinct from P. In my view, Reid's conception of sensation is a bit odd—it suggests that sensations might be reflexive in some manner. I believe that this suggestion is irrelevant to Reid's intention—he simply wants to assert that perception is about external objects but is derived from sensory states that carry no information about external objects. My definitions are intended to capture something like this distinction. My version is more demanding for *perception* (and by implication, less demanding for sensation (which I call "mere sensing"). For a subject S to perceive is for S to have a sense-based impression P of something distinct from S (not merely distinct from P itself).
3 Aristotle's doctrine is well documented in Coren (2019). For an excellent discussion of the medieval tradition, see Christina Thomsen Thörnqvist (2022).
4 For information and discussion about the Indian traditions, I am grateful to my colleagues, Elisa Freschi, Nilanjan Das and Jonardon Ganeri, as well as to Piotr Balcerowicz, who was drawn into the discussion by Jonardon.
5 I hasten to say that I do not know whether this was Jinabhadra's position. It is, however, an undeniable tendency throughout the history of philosophy.
6 Some would respond to this by saying that there are more senses than five—see, for example, Fiona Macpherson (2011). She would say that gravisensing is just an example of an "extra" (as we might say) sense modality. Macpherson is certainly correct to say that there are sense modalities over and above the five (for example, the infrared "vision" that some snakes possess). But her list—"hunger, thirst, wet and dry, the weight of objects, fullness of the bladder, suffocation and respiration, sexual appetite, and lactiferousness" (ibid., 126)—contains a number of special purpose sensory capacities that regulate the body without an associated impression. These are not really full-fledged "senses."
7 Bonnet's work referred to by Moulton, Oliveti and Goriely (2020).
8 Darwin cites a number of other plant physiologists: Hugo De Vries, Sydney Howard Vines, A. B. Frank and Julius von Sachs. Forty years after Darwin, Jagdish Chandra Bose, a Bengali physicist, studied electrical signalling in *Mimosa pudica*. Every Indian child has taken delight in this plant's habit of closing its leaves when touched. Darwin notes that this behaviour of *M. pudica* resembles sleep but says that the cause is different; this in itself indicates this behaviour is actively managed. (See Trewavas (2014, 16–18) for an account of Bose's work and Samuels (2016) for a story in the *Independent* on the Google Doodle that commemorates Bose.)
9 The same could be said, for example, about the way ants use the polarization of the Sun's light to determine direction and bearing. The navigational states of these animals depend on the states of their polarization detectors; however, this is *not* tantamount to representing the Sun or its light as objects distinct from themselves. (Thanks to an anonymous referee for requesting this clarification.)
10 The body schema is implicit in dorsal stream visual processing, or what I have called "motion-guiding vision" (Matthen 2005, ch. 13).
11 The term "cognitive map" was introduced by Tolman, Ritchie and Kalish (1946) and Tolman (1948) and made canonical by O'Keefe and Nadel's (1978) classic work on the hippocampus.

12 Some "lower" animals, such as hymenopterans, move about their habitat, but *navigate* (or move relative to a target) primarily to a home base (such as a nest). It is somewhat controversial whether such navigation demands a full cognitive map or spatial layout, as opposed to a single homing vector. See Wang et al. (2023) for discussion and for compelling evidence that honey-bees possess a full cognitive map. However this may be, it is clear that these animals have sensorimotor othering impressions. For they pick up and transport food from distant locations to their nest.

13 See Rodriguez et al. (2021). Rodriguez suggests that teleost fish have a homologue of the hippocampus.

14 For a discussion of hunger along these lines, see Matthen (2023).

15 Let's say, for the sake of simplicity, that genotype A is more similar to B than to C if it takes fewer mutations to get from A to B than from A to C.

16 See Massa and Gilroy (2003) suggest that pressure on the root deactivates the gravisensors and thus causes the root to grow along the barrier. Of course, this doesn't ensure that the root will emerge on the other side of the barrier relative to the nutrition or wetness gradients.

17 See Tolman (1948) for discussion.

18 May Britt Moser's (2014) Nobel Prize Lecture gives a good and accessible narrative account of these discoveries.

References

Affifi, Ramsey. 2013. "Learning Plants: Semiosis between the Parts and the Whole." *Biosemiotics* 6: 547–59.

Bastien, Renaud, Stéphane Douady, and Bruno Moulia. 2015. "A Unified Model of Shoot Tropism in Plants: Photo-, Gravi-, and Proprio-ception." *PLoS Computational Biology* 11 (2): e1004037.

Calvo, Paco, and Karl Friston. 2017. "Predicting Green: Really Radical (Plant) Predictive Processing." *Journal of the Royal Society Interface* 14 (131): 20170096. http://dx.doi.org/10.1098/rsif.2017.0096.

Coren, Daniel. 2019. "Aristotle on Self-change in Plants." *Rhizomata* 7 (1): 1–30.

Darwin, Charles assisted by Francis Darwin. 1880. *The Power of Movement in Plants*. London: John Murray.

Elster, Jon. 1979. *Ulysses and the Sirens: Studies in Rationality and Irrationality.* Cambridge: Cambridge University Press.

Engelberth, Jürgen. 2003. "Mechanosensing and signaltransduction in tendrils." *Advances in Space Research* 32 (8): 1611–19.

Findly, Ellison Banks. 2002. "Borderline Beings: Plant Possibilities in Early Buddhism." *Journal of the American Oriental Society* 122 (2): 252–63.

Gagliano, Monica, Michael Renton, Martial Depczynski, and Stefano Mancuso. 2014. "Experience Teaches Plants to Learn Faster and Forget Slower in Environments Where it Matters." *Oecologia* 175: 63–72.

Gallagher, Shaun. 1986. "Body Image and Body Schema: A Conceptual Clarification." *Journal of Mind and Behavior* 7 (4): 541–54.

Gallagher, Shaun. 2001. "Dimensions of Embodiment: Body Image and Body Schema in Medical Contexts." In *Handbook of Phenomenology and Medicine* (=Philosophy and Medicine, vol. 68), edited by S. Kay Toombs, 147–75. Dordrecht: Springer.

Gallistel, C. Randy. 1990. *The Organization of Learning*. Cambridge, MA: MIT Press.

Gauthier, David. 1983. "Review of Elster (1979)." *Canadian Journal of Philosophy* 13 (1): 133–40.
Gilroy, Simon. 2008. "Plant Tropisms." *Current Biology* 18 (7): R275–7.
Jaffe, M. J. 1977. "Experimental Separation of Sensory and Motor Functions in Pea Tendrils." *Science* 195 (4274): 191–2.
Landau, Barbara, and Lila Gleitman. 1985. *Language and Experience: Evidence from the Blind Child*. Cambridge, MA: Harvard University Press.
Macpherson, Fiona 2011. "Taxonomizing the -Senses." *Philosophical Studies* 153 (1): 123–42.
Massa, Gioia D., and Simon Gilroy. 2003. "Touch Modulates Gravity Sensing to Regulate the Primary Roots of *Arabidosis Thaliana*." *The Plant Journal* 33 (3): 435–45.
Matthen, Mohan. 1988. "Biological Functions and Perceptual Content." *Journal of Philosophy* 85 (1): 5–27.
Matthen, Mohan. 2005. *Seeing, Doing, and Knowing: A Philosophical Theory of Sense Perception*. Oxford: Clarendon Press.
Matthen, Mohan. 2023. "Hunger, Homeostasis, and Desire." *Mind and Language* 2023: 1–18. https://doi.org/10.1111/mila.12496.
Moser, May-Britt. 2014. "Grid-Cells, Place-Cells and Memory." *Nobel Lecture* 2014. https://www.nobelprize.org/prizes/medicine/2014/may-britt-moser/lecture/.
Moulton, Derek E., Hadrien Oliveti, and Alain Goriely. 2020. "Multiscale integration of environmental stimuli in plant tropism produces complex behaviors," *PNAS* 117 (51): 32226–37.
O'Keefe, John-, and Lynn Nadel. 1978. *The Hippocampus as a Cognitive Map*. Oxford: Oxford University Press.
Pavlovič, Andrej, Viktor Demko, and Ján Hudák. 2010. "Trap Closure and Prey Retention in Venus Flytrap (*Dionaea muscipula*) Temporarily Reduces Photosynthesis and Stimulates Respiration." *Annals of Botany* 105 (1): 37–44. https://doi.org/10.1093/aob/mcp269
Reid, Thomas. 1785/2002. *Essays on the Intellectual Powers of Man*. Edited by Derek Brookes. University Park: Pennsylvania State University Press.
Rescorla, Michael. 2018. "Maps in the Head?" In *The Routledge Handbook of the Philosophy of Animal Minds*, edited by Kristin Andrews and Jacob Beck, 34–45. London: Routledge.
Rivière, Mathieu, and Yasmine Meroz. 2023. "Plants Sum and Subtract Stimuli Over Different Timescales." *PNAS* 120 (42): e2306655120 https://doi.org/10.1073/pnas.2306655120.
Rodríguez, Fernando, Blanca Quintero, Lucas Amores, David Madrid, Carmen Salas-Peña, and Cosme Salas. 2021. "Spatial Cognition in Teleost Fish: Strategies and Mechanisms." *Animals (Basel)* 11(8): 2271 https://doi.org/10.3390/ani11082271.
Samuels, Gabriel. 2016. „Jagadish Chandra Bose: Five Facts You Need to Know about One of the World's Greatest Scientists." *Independent*, November 29, 2016. http://tinyurl.com/gn6p2r4.
Thörnqvist, Christina Thomsen. 2022. "Affected by the Matter: The Question of Plant Perception in the Medieval Latin Tradition on *De somno et vigilia*." In *Forms of Representation in the Aristotelian Tradition. Volume One: Sense Perception*, edited by Juhana Toivanen, 183–212. Leiden: Brill.
Tolman, Edward C. 1948. "Cognitive Maps in Rats and Men." *Psychological Review* 55 (4): 189–208.

Tolman, Edward C., B. F. Ritchie, and D. Kalish. 1946. "Studies in Spatial Learning I: Orientation and the Short Cut." *Journal of Experimental Psychology* 36 (1): 13–24.

Trewavas, Anthony. 2014. *Plant Behaviour and Intelligence.* Oxford: Oxford University Press.

Tye, Michael. 2021. "Qualia." In *Stanford Encyclopedia of Philosophy*, edited by Edward N. Zalta. https://plato.stanford.edu/archives/fall2021/entries/qualia/.

Wang, Zhengwei, Xiuxian Chen, Frank Becker, Uwe Greggers, Stefan Walter, Marleen Werner, Charles R. Gallistel, and Randolf Menzel. 2023. "Honey Bees Infer Source Location from the Dances of Returning Foragers." *PNAS* 120 (12): e2213068120 https://doi.org/10.1073/pnas.2213068120.

Part 3

Learning, Behavior, Affordances

6 No Brain? No Problem

Towards an Ecological Comparative Psychology

P. Adrian Frazier

6.1 Introduction

Plant psychology is a wide open field, with many questions yet unanswered or even unexplored. It presents an opportunity to re-examine old issues in a new light, to explore psychology from a much more fundamental level than before (Colaço 2022). And yet plant psychology often finds itself adopting a classical perspective, a cognitive- and neuro-science developed with humans and, to a lesser extent, animals as its subjects (see Calvo 2016 for a discussion). The logic of this 'classical' viewpoint leads to a number of intractable problems, some of which have been known since the pre-Socratics (Frazier 2021). With the classical cannon of psychological functions comes a temptation to dabble in its more contentious and ill-defined entries, like intelligence, consciousness and even sentience (e.g. Segundo-Ortin and Calvo 2023 and its responses; Segundo-Ortin and Calvo 2021; Trewavas 2016).

Though plant psychology gets considerable attention and has a widening academic literature, the field is still underdeveloped. Few empirical studies have been published asking specifically psychological questions, and even then, the methodology can be deeply flawed (Ponkshe et al. 2023; Rehm and Gradmann 2010; Taiz et al. 2019). The field's presupposed cognitivism and neurobiology provide little to constrain modelling, functioning in some ways more as permission to imagine than as a guide to discovery. And potentially promising approaches like the root brain theory (Baluška et al. 2010) get distorted by the few constraints they *do* suggest, with inappropriately literal translations from the machinery of animal intelligence to that of plants (Rehm and Gradmann 2010; Taiz et al. 2019).

Ecological psychology offers an approach to plants which is simultaneously more radical and more conservative. Radical because it rejects the input processing assumptions of cognitivism. Conservative because it can be limited in scope to questions about how plants do the things they do. And if we need a psychology beyond that, there exists a principled way

DOI: 10.4324/9781003393375-10

to get there, using the conceptual tools of ecological psychology (Gibson 1966, 1979/2014) and interactivism (Bickhard 1980; Bickhard 2009; Bickhard and Richie 1983; Frazier 2021).

In the pages to follow, I will make a case for ecological psychology's theory of perception and action as a candidate model for comparative psychology. The best model on offer currently puts the animal's brain and its evolution at the centre of analysis. Whatever its value, that model necessarily excludes plants. Even *if* plants have 'nervous systems', they have little or no evolutionary history in common with those of animals. On the other hand, organisms of all kingdoms and all domains engage with *affordances*, approaching, moving between, grasping and so on, the surfaces of their environment. According to ecological psychology's hypotheses, the composition and layout of affordances is what makes them approachable, move between-able, graspable and so on. And the same composition and layout structures light and other ambient energy fields in a way that makes them detectable *as* approachable, etc. and informs their activities (Gibson 1979/2014). Information of this sort, then, functions as a selection constraint on the evolution of control processes, and affordances function as constraints on behaviour. This is true of evolution, development, learning and action alike.

The chapter will proceed as follows. Before diving into ecological psychology, I will discuss the 'brain' framework, with respect to animals and plants; what it offers and what it does not. I will follow this with an introduction to ecological psychology, with special focus on affordances and information, along with their 'dual' concepts, effectivities and control. I will explore a pair of examples from the world of climbing plants and demonstrate how an ecological psychologist might approach them.

6.2 Even *with* Brain, Only So Much Gain

The brain-centric study of psychology has the advantage of a comparative science based on the developmental and evolutionary relationships between different animals' brains. The amygdala, for instance, is shared by all tetrapods (four-legged animals), emerging in evolution as part of the olfactory bulb (Pabba 2013). Knowing about functions served by the human amygdala gives us reason to suspect that the same functions are served by those of other tetrapods. On the other hand, the organisation and connectivity of amygdalae differs between mammals and their reptile and anuran ancestors. These differences, and differences in behaviour, can then be investigated to recalibrate our understanding of the amygdala's functions and their evolution. In general, if you have a theory about what some organ of the brain does, and you find the same organ in the brains of other creatures, then you have grounds for comparison (cf., Feinberg and Mallatt 2016).

None of the foregoing is possible when crossing kingdoms. Even if plants have something sensibly called a 'nervous system', it has no evolutionary relationship to the nervous systems of animals. Even among animals, there appears to be more than one lineage (i.e., the ctenophore 'nerve net'; Jager and Manuel 2016). What plants *do* share with animals, such as neurotransmitter substances, ion channels and cells with action potentials, are not organised into the kinds of organs found in animals—as far as we know. Action potentials *are* involved in rapid movements, but the cells realising them are not organised as nets, much less nuclei and the organs they make up in animals. And why would they be? Most plants get along fine at *very* slow timescales, and they need not undergo the rapid qualitative transformations in behaviour characteristic of animals with complicated brains (Fultot et al. 2019).

Suppose, however, that plants have evolved a different kind of 'nervous system', with entirely different topology and composition. Plants have highly complicated transduction systems, with numerous kinds of pathways. The 'nuclei' of such systems might be distributed over the roots and shoots, with various signalling arrangements realising a recursive and integrative architecture, with psychological functions emergent in their organisation. Even so, how do we know that these functions are anything like those of the animal's repertoire? Looking for something like 'consciousness' in plants might be like searching under a street lamp for something lost in the dark.

On the other hand, perhaps the plant's nervous system evolved similar functions to those of animals by convergent evolution. Without a theory, however, we are left to guess which functions we ought to look for. We cannot rely on shared evolutionary history. And we cannot look for homologous structures. The best we can do is compare behaviour. But then, all we have are data and our intuitions about what they mean. In itself, 'nervous system' as the ground for comparison is shaky ground indeed.

6.2.1 An Alternative?

Convergent evolution implies that there exists something in common about the lifestyles of the convergently evolved; common opportunities, challenges and constraints. If members of the two kingdoms can *do* the same things in the same kinds of environments, then according to ecological psychology, we should expect for those environments to provide the same information to the plant, the animal or any other creature, about the *availability of the doing* and *how to do it*. The opportunity to do the thing, along with the need to coordinate activities around the information for doing so, serve as common constraints on action, learning, development and evolution—for *all* creatures (at least those capable, potentially or actually, of doing the thing).

In the next sections, I will focus primarily on ecological psychology and the main elements of its theory of perception: affordances and information pickup. We may not need anything more than these concepts to fully understand the plant's psychology. Although the theory is radical, the methods are additive. We are looking for variables similar to some of those already identified in the physiological literature, but a little bit richer, a bit more dynamic. And the upshot is a principled way to approach plant psychology, however much or little we need of it in the end.

6.3 Ecological Psychology

Ecological psychology is, primarily, a theory of perception and action developed by James J. Gibson (1966, 1979/2014). The bulk of Gibson's writing concerned visual perception, but his theory was meant for all sensory modalities. His ideas developed around criticism of sensation-based theories of perception, or 'inputs-followed-by-processing-followed-by-perception' models (Bickhard and Richie 1983, 11).

Gibson developed some of his 'ecological' insights early on, taking his starting point to be the animal steering its locomotion through a cluttered environment. He searched for concepts that would be common to locomotion of any sort, focusing on things like the terrain, obstacles, collisions and paths (Gibson and Crooks 1938; Reed 1988). Though he did not write about it at the time, he had intuitions that the point of reference is the animal itself and that it must involve symmetricalising, minimising or maximising, or centring something (Reed 1988). This insight was the germ of his later theory of information pickup. When interacting with the environment, he would argue, the animal 'picks up' invariants and disruptions in the structure of light and other stimulus fields, which constitutes 'information about' the composition and layout of the environment—and ultimately, its affordances (Gibson 1966, 1979/2014).

Gibson's theory of information pickup was preceded by what is sometimes called an 'ecological direct encoding' theory (Bickhard and Richie 1983). He looked to the environment-organism relationship for inspiration, but it was variables of the retinal image he took to furnish information about the environment, implicating a direct correspondence between retinal image and perception (Bickhard and Richie 1983; Gibson 1950). His information pickup theory has no such implication.[1] Gibson abandoned the retinal image for ambient light, the structure of which results from its interactions with surrounding surfaces. The light, then, carries information in the sense of 'correlates with' the composition and layout of surfaces. The animal, however, does not interact with light looking for correlations, not as such. It interacts with light or anything else for its effect on the course of interaction. These effects, then, the internal states they yield, are exploited by the animal's brain, or more generally, the organism's control processes. Sensory stimulation, then, is *elicited* rather than imposed. Let us explore the concept of 'perception' at work here.

6.3.1 Perception

In Gibson's theory of perception, perceiving *just is* detecting information (Michaels and Carello 1981). For a moment, I will step away from that language and use instead the language of *interaction*, which can be understood as follows: An organism performs an action, and the course of doing so is influenced by the environment (Frazier 2021; cf., Bickhard 1980). Perception, in this formulation, *just is* interaction. The result of an interaction implicitly differentiates one environment from the other, since different environments influence action in different ways. When an organism engages in contact with the environment, with its surfaces, with the light reflecting from them, and so on, it yields sensor stimulations and reaction forces, among other things. It is by these effects that the organism picks up and exploits the structure of surrounding stimulus fields.

Consider bacterial chemotaxis as an example (Webre, Wolanin, and Stock 2003). The environment is filled with diffusing chemical compounds, some being helpful 'attractants' and others being harmful 'repellents'. To reach the source of an attractant, bacteria enact a biased random walk, swimming if moving up gradient and tumbling to reorient otherwise. The swimming is accomplished by an array of flagella rotating counter-clockwise together. If any of them starts rotating clockwise, the whole array gets jammed up and the bacterium starts tumbling. The clockwise rotation is caused by the action of phosphorylated CheY, which is increased in concentration by the auto-phosphorylation of another protein, CheA. This in turn is (in effect) 'turned off' when attractants stimulate the sensors. Keeping it turned off, however, requires increasing frequency of stimulation, which means swimming up gradient. The attractant alone is not sufficient to keep the bacterium swimming, and it would be uninformative in any case. Only as a *gradient*, and only when *picked up as a gradient*, is the substance informative about when to swim and when to tumble.

The foregoing example illustrates some of what is meant by interaction and what interacting does. I have already suggested that perceiving is interacting. As such, perceiving yields two outcomes: (A) differentiations, and (B) the set-up of control flows in accordance with what was differentiated. The first of these is factual, the raw outcome of interacting with the environment. The second has truth value. The organisation of control processes implicitly presupposes something about the environment—*which might be false*. The bacterium could be swimming up a saccharine gradient, which has the same effects on sensors as sugar but offers none of the nourishment. This point will not factor much in the discussion to follow, but it serves as the groundwork for an interactive theory of representation and, by extension, cognition, which is compatible with ecological psychology and its contention that perception is the source of knowledge (Bickhard 1980, 2009; Bickhard and Richie 1983).

6.3.2 Environment and Information

In the appendix of his 1979/2014 book, Gibson defined 'environment' in terms of surfaces, substances, and the medium. A surface is the outermost part of an object, that from which light and pressure waves reflect and the source of chemical diffusion. Substance is the composition of an object, the stuff below the surface. And the medium is the space between surfaces. Surfaces have an arrangement, which is carried in the structure of ambient light and pressure waves, diffusion gradients and so on. As is the composition of surfaces. Said differently, the structure of ambient light (etc.) carries *information about* the composition and layout of environments in the sense of being correlated with it. For instance, the angle between two observable points from a point of observation increases exponentially with proximity (the angle is given by the inverse tangent, thus exponential for a constant rate of movement). The array of light (etc.) reflected (etc.) from the surfaces in view expand likewise, closer surfaces expanding more rapidly than further ones. By interacting with the light (etc.) the organism picks up its correlation with the environment's composition and layout. That is, interactions implicitly differentiate one environment or situation from another (Bickhard 2009). If the action is 'moving forward', then a symmetrically expanding field differentiates a situation in which continuing to move forward will get the organism closer to the surface from one in which doing so *will not*.

As will be explicated below, there exists a curiously reciprocal relationship between all these concepts: environment, organism, perception, action, information and control. Before getting there, however, we need to add another pair of terms to the pile, 'affordances' and 'effectivities'. The first of these terms, affordances, is Gibson's best-known theoretical contribution, though considerable disagreement exists about what he meant by it and how it ought to be conceptualised today. Most differences can be attributed to interpretations of 'organism-environment complementarity', the most notable other than that which I will discuss below being Chemero (2009), Reed (1996), Stoffregen (2003) and Turvey (1992).

6.3.3 Affordances

In his 1979/2014 book, Gibson gave his most detailed explication of 'affordances', which he defined as follows: 'The *affordances* of the environment are what it *offers* the animal, what it *provides* or *furnishes*, either for good or ill' (Gibson 1979/2014, 19, emphasis original). At that point in the book, Gibson had claimed that animals perceive surfaces and that they perceive affordances. His argument for the former is that ambient light is structured in a manner specific to the composition and layout of the environment. But how, Gibson asked, do we go from surfaces to affordances? As an answer, he suggested that the composition and layout of an environment *constitutes* what it affords. This point is especially obvious in the case of a surface being *approachable*. A tree that is too far away or blocked by impassable obstacles obviously cannot be approached. Indeed, it may not even be visible. Clearly an aspect of the layout. Other things about trees make them good hosts to vines, like *Monstera tenuis*, some of which are specific to their composition: tall enough, wide enough, not already covered in greenery and so on. These same features make trees detectable *as* climbable, because of their effect on the light.

Gibson's definition, however, leaves a bit of ambiguity. An apple tree offers apples to apple-eaters, but should we consider apples to be affordances? He provided an example as a clue to what he meant: "If a terrestrial surface is nearly horizontal [...], nearly flat [...], and if its substance is rigid (relative to the weight of the animal), then the surface *affords support*" (Gibson 1979/2014, 119, emphasis original). Stated differently, the surface functions as support for certain actions, like stand, walk or jump. Thus, we say the surface is stand-able, walk-able or jump-able. These '-ables' are the surface's *affordances*. The organism side of this, the standing, walk-ing or jump-ing, are the organism's *effectivities*.

What about the apple tree? On the one hand, apples themselves have affordances, being pickable, edible and so on. And apple trees offer access to them. A construction relating apple trees and apples in terms of functions served to an apple-eater's food seeking, picking, eating, etc. is surely possible, so in that sense it is true that 'apples are affordances of trees'. It also makes sense to relate apples and apple trees in terms of their layout and composition. But then again, by hypothesis, the composition and layout *constitute* what the surfaces afford.

The foregoing line of argumentation may seem like running a loop, and it should. The environment-organism construction and affordance-effectivity construction are *dual descriptions* of an ecosystem. As is each member of the two pairs—at least according to the intentional dynamics framework in ecological psychology. Duality, as a framework for modelling and problem-solving in general (Karasik 2021), could be particularly useful for building a well-grounded comparative psychology.

Duality

Duality implicates a symmetry of perspective; one thing can be looked at from two equivalent points of view. Consider the following statement about points and straight lines on the projective plane: 'Every two straight lines have one and only one common point' (borrowed from Karasik 2021, 10). The statement is equally true if we swap 'straight line' and 'point'. Identifying straight lines and points as being dual objects, with respect to projective planes, means that theorems in the language of points are dual to theorems in the language of lines and vice versa. In general, changing perspectives by transitioning from one side of a dual to another or from one pair of duals to another can be tremendously useful, making the solution to an intransigent problem clearer and easier to find. Duality is one of the most important developments in theoretical inquiry, with innumerable applications in physics and mathematics, among other things (Karasik 2021).

Shaw and Turvey (1981) proposed a duality framework for studying 'ecosystems', relational structures with four pairs of duals: Environment and organism, affordance and effectivity, perception and action, and information and control (see also Shaw and Kinsella-Shaw 1988; Shaw and Kinsella-Shaw 2012). Each of these constitutes a finer grain of analysis (respectively), and each is mutually constraining of each other. To understand how this works, let us begin with the coarsest grain, environment and organism. One way to understand 'environment' is in Gibson's terms, as the surfaces, substances and medium of an organism's surrounds (being lenient to some degree about what these might mean at microscopic scales). We might specify a particular surface or equivalence class of surfaces, like 'apple'. By duality, this picks out an equivalence class of organisms for whom 'apple' is part of the environment. The same is true for *anything else* we might pick out as 'environment'. Now, we can turn this around. Any organism or equivalence class of organisms we pick out defines, by duality, an equivalence class of environments.

Zooming back out for a moment, we can think about an 'ecosystem' as one thing, and then zooming back in, 'environment' and 'organism' are *aspects* of an ecosystem, *dual* aspects. As a grain, 'environment and organism' is, in itself, another aspect, as are the other three pairs. The term 'aspect' here, like 'duality', means something akin to 'perspective', but of a more general kind than duality. Are any or all of the four pairs, *dual* aspects? Perhaps, but if so, only under the condition of mutual constraint. Duality is a kind of isomorphism, meaning that there must be a one-to-one mapping from one side of the dual to the other.

For an example, consider once again 'apple' as the environment. Apples have many affordances for many different kinds of creatures. This makes sense, given that the level of affordance and effectivity is a finer grain than that of environment and organism. However, the one-to-many mapping here already rules out duality between apples and their affordances. Suppose we go further, to perception and action. If we shift perspective to action instead of environment, we might specify 'throw'. One thing that makes this level useful is that we can be more specific about affordance and effectivity: 'throw-able' and 'throw-ing' respectively. We can do the same for environment and organism: 'throw-ee' and 'throw-er'. Apples, now, are elements of an equivalence class of objects 'throw-ees' with the affordance 'throw-able'. The labelling scheme breaks down, unfortunately, for 'perception', 'information' and 'control', but it helps to illustrate my point. By singling out 'apple' alone, 'throw' is just one action among many; and by singling out 'throw' alone, 'apple' is just one environment among many. When they constrain each other, the equivalence class of ecosystems gets much smaller.

The finest grain, information and control, is in my opinion the most important. An action cannot be realised with reliable success without control processes 'tuned-in' to information specific to it, at least in its local context—if the organism were to wander out of that context, then it would need a more general variable. The control processes involved in the *Monstera tenuis* seedling's journey to its host tree may be tuned only to the darkest sector of its horizon, for instance (see below). But there might be environments (other than experimental situations) in which the 'darkest sector' alone does not specify where to grow. If getting to host trees still matters in those environments, then the vine's seedlings will be unprepared to do so.

Here again we find mutual constraint. By specifying information, we limit action to 'throw'. And even though we are now limited to being a throw-er throw-ing a throw-able throw-ee, it is not clear that we have locked-in the needed degrees of freedom. The same throw-er can throw in a lot of different ways. And 'throw-ing' can, generically, be realised by many different kinds of throw-ers. This requires that we shift back up to the level of environment and organism. Using a well-known example from the ecological psychology literature, consider 'stepping'. Warren (1984) demonstrated that to perceive a stair as step-able, the stair's riser height must have the right proportion to the step-er's leg. Throwing and grasping, similarly, require a throw-able or grasp-able object with the right proportion to the thrower's or grasper's hand (or whatever they use to throw or grasp). Whether this move takes us to the needed degree of specificity to reach mutual duality between all four pairs is unclear to me, but it illustrates my point. Each move we make with one element of the 'ecosystem' constrains the moves we can make with others.

One way to approach this framework is to treat the fully or maximally dualised 'ecosystem' as a kind of 'spherical cow'. It provides a normative model that we can use to understand not only how plants do what they do, but also how they get there, whether by evolution, development or learning. Observing that a vine seedling grows directly towards its host tree tells you at least that 'approach-ing' is part of the equation. If so, then we already know something about the tree's affordance 'approach-able', and with that, we get a hypothesis for free about how it does so: by moving so as to expand the optical array. If this hypothesis turns out to be false, then we still have grounds to explore further. Perhaps the hypothesis is true for some other vines which also grow directly to their host. From there, we can ask questions about the kinds of environments each inhabits. Do the latter live in more variable environments and therefore require greater informational specificity?

6.3.4 What about Plants?

Gibson's theory was developed with terrestrial animals in mind, and he was doubtful about anything beyond that. In fact, he explicitly rejected the idea that his theory had anything to say about plants (Gibson 1979/2014). Part of the reason for this is that he was conservative in that he was not willing to extend his ideas beyond that which he had already worked out himself. He had even expressed doubt about whether his scheme would work under water (William Mace, personal communication). Those who followed him, however, have not been so conservative.

Even at the time Reed and Jones (1977) wrote about a then-recently discovered tropical vine, *Monstera tenuis*, which had been shown to grow directly towards its host tree as a seedling, seemingly doing so according to an optical variable. In this example, Reed and Jones saw Gibson's principles at work. Several decades later, Claudia Carello and her coauthors (2012) argued for plants as an example of intelligence *without* a nervous system, rejecting the concept of 'plant neurobiology' in favour of an approach based in ecological psychology. Michael Turvey and his collaborators have been especially keen to push the boundaries of ecological psychology, with calls for a theory of perception and action for all creatures (Turvey 2018), a theory of intelligence from first principles (Turvey and Carello 2012) and a view of biology and psychology based on the thermodynamics of dissipative structures (Kugler, Kelso, and Turvey 1980) and entropy production (Swenson and Turvey 1991).

Unfortunately, the question of whether we ought to take Gibson's framework into the greenhouse is still largely untested. Plant psychology itself is still a mostly untested concept. Nevertheless, we can draw from existing literature for inspiration. Gibson probably did not know about the *Monstera tenuis* vine Reed and Jones (1977) wrote about, and the plant sciences have advanced considerably since that time. It is fitting, then, to revisit that case and compare its behaviour with that of another vine with very different behaviour.

6.4 A Tale of Two Vines

The host-finding behaviour of vines is an especially interesting case, charming researchers since at least the time of Charles Darwin (C. Darwin 1875; C. Darwin and F. Darwin 1880). In the abstract, the vine's lifestyle is a kind of shade-avoidance strategy. *Monstera tenuis*, in particular, starts life under an especially dense canopy, and climbing the nearest tree is the only way to escape the dark and access light. Instead of competing with trees, vines commit their resources to growth, oftentimes spreading themselves over whole surfaces. This they do instead of growing a supportive trunk, like trees. Vines, then, have greater flexibility and motility, some abandoning roots altogether, living most of their lives colonising other plants.

In this section, I will consider a pair of vines with diametrically opposed host-finding strategies. In either case, the affordance for individual approaching and climbing exists, but the size and distribution of potential host trees do not afford the same to groups. I present these cases not necessarily as evidence in favour of ecological plant psychology but to demonstrate how an ecological psychologist would approach them, to raise the kind of questions an ecological psychologist would think to ask.

6.4.1 Monstera tenuis

The host-finding behaviour of the tropical vine *Monstera tenuis*, formerly *M. gigantea*, was first reported by Donald Strong and Thomas Ray in 1975. The vine's seedlings typically germinate in very low light conditions, under a dense, dark leaf canopy in the forests of Central America (Strong and Ray 1975). The seedling lives on seed reserves while searching for a host tree, but it only lasts for 2 to 3 m of growth. Fortunately, the seedlings are dropped by their parent plant from the very tree that they will soon occupy, so long as they survive the trek. But even so, no time can be wasted on random search and dead ends.

Germination begins almost immediately after reaching the ground. The seedling begins its trek by standing up, bending towards the tree, and growing. Strong and Ray (1975) proposed *skototropism*, or growth into the dark, to explain how the seedling accomplishes this feat. Specifically, it grows in the direction of the darkest sector of its horizon. To rule out negative phototropism, or growth away from the *lightest* sector, Strong and Ray took measurements of *M. tenuis* seedlings' angle of growth relative to the tree and took an average for each. While the path is basically direct, debris and other obstacles can cause it to be contorted. Seedlings within 2 m of the tree grew at an average angle of 0°. Average angle increased with distance, approaching 135° in some cases. Standard deviation was lower for seedlings within 1 m, and 2 to 3 times more variable beyond that (based on the data presented in their Figure 2, only one of the 19 trees sampled). Importantly, the average angle was never 180° from the lightest sector.

In a series of experiments with fake 'trees' made of dark panels, Strong and Ray (1975) established several parameters involved in *M. tenuis'* detection and migration to the host tree. The researchers set seedlings to grow parallel with the panels. Those that detected the panels changed course entirely, turning 90° towards them. Only around half of seedlings within 30 cm of 40% shade panels turned and it took more than 12 days to do so. Further than that, however, none turned. On the other hand, all but the most distant seedlings from 64% shade panels made the turn, closer seedlings turning within 6 days, while more distal seedlings took longer. Results were no different for darker shade panels.

The results described above provide evidence that growth is towards dark and not away from light. And that somewhere between 40% and 64% shade, a surface is dark enough to detect. But without an experiment offering seedlings a choice of shade panels, we cannot be sure that the seedlings grow towards the *darkest* sector of horizon. Might there be a better explanation? A related, but not identical variable? Recall that seedlings growing further from the panels took longer to change direction, and seedlings further from the trees diverged from 0°, sometimes considerably, and with greater variability.

In another experiment, Strong and Ray (1975) demonstrated that the panels should take up at least 21° of the horizon to be reliably detected. They placed seedlings at various distances from dark panels of widths between 1 and 30 cm (growing parallel, once again). The wider the panel, the more distal the seedlings it could attract. The authors did not provide detailed data, but based on what information they *did* provide, the minimum angle for reliable detection appears to be about 21° for wide and narrow 'trees' alike. This was true also of the tree they used as an example in their Figure 2, which was 75 cm in diameter. At 2 m, its sector angle was about 21°; at 1 m, the tree's sector angle is almost twice that, at about 41°.

Using the same calculations, we can model expansion and contraction of the sector as the opposite angle to the surface. For a constant rate of movement towards the surface, the angle increases *exponentially*. This means that in close proximity, the angle increases dramatically with each unit of movement. But with some distance, the angle increases only slightly with each unit of movement. Thus, if the sector is detected at all, doing so would take longer than it would closer to the surface: precisely that which Strong and Ray (1975) observed.

A question can be asked about just how tree trunks structure light, such that they can form a detectable sector of the seedling's horizon. Strong and Ray (1975) characterised their sector as 'darkest', but what makes something 'dark' from a plant's perspective? In the plant world, there are two kinds. 'Pseudo-dark' conditions are those plants find themselves in when shaded by other plants. Reflectance from other plants is low in blue and red light but high in green and far-red. In 'true dark' conditions, there is very little if any light at any wavelength. Tree trunks reflect very little light, but not necessarily less than surrounding foliage at the same height under canopy (Endler 1993). Of course, there is very little foliage under the canopy, so light reflected from surfaces at the same height as a tree trunk would mainly be other tree trunks further away. The nearest tree trunk would be 'brighter' since its reflectance would take up more of the horizon. We might be better off conceptualising the horizon as a sphere of light and shade sectors. The lower half of this sphere would be consistently dark when a seedling grows along the ground. The upper half would be constituted in part by light transmitted through canopy leaves (and the occasional speck between them), and its radiance would be greater in magnitude than that of a tree trunk, and it would have much more green than red light—trunks reflect more or less the same amount of all wavelengths from green to red. A nearby tree trunk would take up a full wedge of the spherical horizon. The seedling, then, could differentiate a potential host tree from the ground by the fact that the tree's wedge increases or decreases in width with different movements, and the ground does not (unless the seedling is growing upward). Growing up a tree trunk would be similar, with the trunk taking up half the spherical horizon.

Whatever the case, the results of Strong and Ray's (1975) measurements give us reason to believe that 'approaching' is not different for *M. tenuis* seedlings than it is for anything else, as far as the information is concerned. If we had data on the 'turning' enacted by Strong and Ray's seedlings, there might be be a stronger case. At whatever distance a seedling grows parallel to a panel, the largest sector angle is attained while approaching and crossing the half-way point. As before, expansion to maximum is much more subtle further from the surface. In looking for control processes, we should expect to find them operating on the effect of expansion and contraction. As an example from another domain, bacterial chemotaxis involves a process which attenuates the effect of a given attractant's stimulation, so that it must be moving up gradient, encountering more and more of the attractant to keep swimming (Webre, Wolanin, and Stock 2003). Approaching, once again, is informed by an expanding portion of the sensory array, chemical, optical or otherwise.

If the information for approaching really is universal, why do some species not make use of it to do their approaching? I will explore this briefly with another example, a genus of vines with a host-seeking phase, but no apparent strategy for doing so. As it turns out, given the habitats they occupy, widespread competence at approaching and selecting between hosts would be detrimental for the whole population.

6.4.2 Heteropsis

Members of the genus *Heteropsis* described by Balcázar-Vargas, Peñuela-Mora et al. (2012) are equally as remarkable as *Monstera tenuis*, but for quite different reasons. Like *M. tenuis*, these vines start life as leafless seedlings in search of a host. They cannot reach sexual maturity without climbing high enough: 1.5 m for *H. oblongifolia* and 10 m for *H. flexuosa* and *H. macrophylla*. Despite the reproductive necessity of finding a suitable host, doing so does not appear to be their first priority. Evidently, they do not exercise any discernment whatsoever, not in seeking, nor in selecting a host.

After germination, the seedlings scatter in all directions, not oriented to light, dark, or anything else (Balcázar-Vargas, Peñuela-Mora et al. 2012). Their growth is tortuous, with arbitrary twists and turns. As soon as they find something to climb, they do so. It could be an adult tree, but it could also be a sapling. It could be a shrub or an herb. It could be a fallen log or a rock. Nothing other than being a climbable surface seems to matter.

After getting off the ground, they send an absorbing root to the soil. Apparently, this step in their development is the first priority. Like *M. tenuis*, *Heteropsis* seedlings live on seed reserves for the first days of their lives, and for *Heteropsis*, failing to drop root severely limits the chance for survival (Balcázar-Vargas, Peñuela-Mora et al. 2012). Once established, the plants can then focus on reproductive concerns. If the present host is too short, they can send clones to search for a better one, but most of them do not. Perhaps this is because only around half will succeed, and even these contribute very little to the overall population dynamics, which is driven primarily by the 19% of seedlings who find a suitable host on the first try (Balcázar-Vargas, Salguero-Gómez, and Zuidema 2015).

If sexual reproduction depends on finding suitably tall trees, how can populations of these vines get away with being, in effect, *blind* to them? Balcázar-Vargas et al. (2015) argue that they exhibit strong *source-sink* dynamics. Populations can persist in habitats poor in the resources necessary for reproduction by functioning as 'sinks', with a flow of migration from 'source' populations from habitats richer in resources. The kind of scattering evidenced by *Heteropsis* seedlings keeps the dynamic in motion. In their case, the resource in question is *host trees* in sufficient quantity and with enough room to support a larger population of vines.

The habitat of *Heteropsis* is sparser than that of *M. tenuis*, and not nearly as dark. Fruiting requires access to direct sunlight, so they must climb. Unlike the large-diameter trees available to *M. tenuis*, the trees of *Heteropsis*' habitat are much narrower by comparison. Most seedlings end up on hosts less than 10 cm in diameter. And upwards of 60–80% of seedlings who make it to adulthood occupy trees between 10 and 40 cm². Recall that the tree represented by Strong and Ray's (1975) Figure 1 was 75 cm, more than twice the diameter in most cases. If *Heteropsis* was as skilled at locating and growing towards host trees as *M. tenuis*, the seedlings would crowd each other out very quickly. The trees of this environment afford approaching and climbing, but only to individuals, not populations or even sub-populations. The vine's host-blindness, then, is quite tuned-in to the affordance structure of its environment.

6.5 Conclusion

The use of psychological concepts in the study of plants is precarious, and it requires great theoretical care if it is to be successful. Whatever plant psychology turns out to be, it may not be nearly as exciting as some might imagine. Though Gibson never saw plants in his theory of perception, the conceptual tools he introduced may be just 'unexciting' enough to do the job, an incremental step for plant psychology. Ecological psychology's theory of information pickup and affordances or the duality framework may be all we need. But as suggested in 'Perception', there exists a principled route to further kinds of knowing and cognition. Explaining it properly would go beyond the scope of this chapter, but for the interested reader, Eleanor J. Gibson's work on development and learning (E. J. Gibson 1988) is an excellent place to start. And interactivism, a theory developed by Mark H. Bickhard, has been extended to many different areas of psychology (Bickhard 1980, 2009; Bickhard and Richie 1983; Frazier 2021).

Acknowledgements

This research is supported by the Office of Naval Research Global Award # N62909-19-1-2015. Thanks to Thomas Ray for comments and information about *Monstera tenuis*.

Notes

1 Criticisms of Gibson's theory usually proceed from the assumption that his later work *does* require a direct correspondence between perception and stimulation and usually does not take his later interactive theory into account (e.g., Fodor and Pylyshyn 1981; Ullman 1980; see Bickhard and Richie 1983 for critical analysis of their arguments). Even Noë (2004), a much more sympathetic figure, gets this wrong, suggesting that Gibson 'held that his 'ecological approach' can handle the problem of inverse optics' which he criticised earlier in the book. 'This problem' Noë continues, 'turns out to be a consequence of the optional assumption that the data for vision is confined to the retinal image' (p.21). Gibson rejected this assertion at the beginning of his 1966 book: 'the input of the sensory nerves is not the basis of perception [...]. These are not the data of perception, not the raw material out of which perception is fashioned by the brain' (p. 5). Noë, like Gibson's critics, is mistaken here. The retinal image theory (the direct encoding theory) was abandoned in favour of the information pickup theory.
2 One of the species observed, *H. Oblongifolia*, is much smaller and needs not climb as high. Its adults were found on hosts less than 10 cm.

References

Balcázar-Vargas, María P., María C. Peñuela-Mora, Tinde R. van Andel, and Pieter A. Zuidema. 2012. "The Quest for a Suitable Host: Size Distributions of Host

Trees and Secondary Hemiepiphytes Search Strategy." *Biotropica* 44 (1): 19–26. https://doi.org/10.1111/j.1744-7429.2011.00767.x.

Balcázar-Vargas, María P., Roberto Salguero-Gómez, and Pieter A Zuidema. 2015. "No Second Chances: Demography from the Forest Floor to the Canopy and Back Again." *Journal of Ecology* 103 (6): 1498–508. https://doi.org/10.1111/1365-2745.12466.

Baluška, František, Stefano Mancuso, Dieter Volkmann, and Peter W. Barlow. 2010. "Root Apex Transition Zone: A Signalling-Response Nexus in the Root." *Trends in Plant Science* 15 (7): 402–8. https://doi.org/10.1016/j.tplants.2010.04.007.

Bickhard, Mark H. 1980. "A Model of Developmental and Psychological Processes." *Genetic Psychology Monographs* 102 (1): 61–116.

Bickhard, Mark H. 2009. "The Interactivist Model." *Synthese* 166 (3): 547–91. https://doi.org/10.1007/s11229-008-9375-x.

Bickhard, Mark H., and D. Michael Richie. 1983. *On the Nature of Representation*. New York: Prager.

Calvo, Paco. 2016. "The Philosophy of Plant Neurobiology: A Manifesto." *Synthese* 193: 1323–43. https://doi.org/10.1007/s11229-016-1040-1.

Carello, Claudia, Daniela Vaz, Julia J. C. Blau, and Stephanie Petrusz. 2012. "Unnerving intelligence." *Ecological Psychology* 24 (3): 241–64. https://doi.org/10.1080/10407413.2012.702628.

Chemero, Anthony P. 2009. *Radical Embodied Cognitive Science*. Cambridge, MA and London: MIT Press.

Colaço, David. 2022. "Why Studying Plant Cognition is Valuable, Even if Plants Aren't Cognitive." *Synthese* 200: 453. https://doi.org/10.1007/s11229-022-03869-7.

Darwin, Charles. 1875. *The Movements and Habits of Climbing Plants*. London: John Murray.

Darwin, Charles assisted by Francis Darwin. 1880. *The Power of Movements in Plants*. London: John Murray.

Endler, John A. 1993. "The Color of Light in Forests and its Implications." *Ecological Monographs* 63 (1): 1–27. https://doi.org/10.2307/2937121.

Feinberg, Todd E., and Jon Mallatt. 2016. "The Nature of Primary Consciousness. A New Synthesis." *Consciousness and Cognition* 43: 113–27. https://doi.org/10.1016/j.concog.2016.05.009.

Fodor, Jerry A., and Zenon W. Pylyshyn. 1981. "How Direct is Visual Perception? Some Reflections on Gibson's 'Ecological Approach'." *Cognition* 9 (2): 139–69. https://doi.org/10.1016/0010-0277(81)90009-3.

Frazier, P. Adrian. 2021. "On the Possibility of Plant Consciousness: A View from Ecointeractivism." *Mind and Matter* 19 (2): 229–59.

Fultot, Martin, P. Adrain Frazier, Michael T. Turvey, and Claudia Carello. 2019. "What Are Nervous Systems For?" *Ecological Psychology* 31 (3): 218–34. https://doi.org/10.1080/10407413.2019.1615205.

Gibson, Eleanor J (1988). 'Exploratory behavior in the development of perceiving, acting, and the acquiring of knowledge'. In: Annual review of psychology 39.1, pp. 1–42.

Gibson, James J. 1950. *The Perception of the Visual World*. Boston, MA: Houghton Mifflin

Gibson, James J. 1979/2014. *The Ecological Approach to Visual Perception: Classic Edition*. New York: Psychology Press. https://doi.org/10.4324/9781315740218.

Gibson, James J. 1966. *The Senses Considered as Perceptual Systems*. New York: Houghton Mifflin.

Gibson, James J., and Laurence E. Crooks. 1938. "A Theoretical Field-analysis of Automobile-Driving." *The American Journal of Psychology* 51 (3): 453–71. https://doi.org/10.2307/1416145.

Jager, Muriel, and Michaël Manuel. 2016. "Ctenophores: An Evolutionary-developmental Perspective." *Current Opinion in Genetics & Development* 39: 85–92. https://doi.org/10.1016/j.gde.2016.05.020.

Karasik, Yevgeny B. 2021. *Duality Revolution*. Independently published.

Kugler, Peter N., J. A. Scott Kelso, and Michael T. Turvey. 1980. "On the Concept of Coordinative Structures as Dissipative Structures: I. Theoretical Lines of Convergence." *Advances in Psychology* 1: 3–47. https://doi.org/10.1016/S0166-4115(08)61936-6.

Michaels, Claire F., and Claudia Carello. 1981. *Direct Perception*. Englewood Cliffs, NJ: Prentice-Hall.

Noë, Alva. 2004. *Action in Perception*. Cambridge, MA: MIT Press.

Pabba, Mohan. 2013. "Evolutionary Development of the Amygdaloid Complex." *Frontiers in Neuroanatomy* 7. https://doi.org/10.3389/fnana.2013.00027.

Ponkshe, Aditya, Jacobo Blancas Barroso, Charles I. Abramson, and Paco Calvo. 2023. "A Case Study of Learning in Plants: Lessons Learned from Pea Plants." *Quarterly Journal of Experimental Psychology* 17470218231203078. https://doi.org/10.1177/17470218231203078.

Reed, Edward S. 1988. *James J. Gibson and the Psychology of Perception*. New Haven: Yale University Press.

Reed, Edward S. 1996. *Encountering the World: Toward an Ecological Psychology*. New York: Oxford University Press.

Reed, Edward S., and Rebecca K. Jones. 1977. "Towards a Definition of Living Systems: A Theory of Ecological Support for Behavior." *Acta Biotheoretica* 26 (3): 153–63. https://doi.org/10.1007/BF00048424.

Rehm, Hubert and Dietrich Gradmann. 2010. "Intelligent Plants or Stupid Studies." *Lab Times* 3: 30–32.

Segundo-Ortin, Miguel, and Paco Calvo. 2021. "Consciousness and Cognition in Plants." *Wiley Interdisciplinary Reviews: Cognitive Science* 13 (2): e1578. https://doi.org/10.1002/wcs.1578.

Segundo-Ortin, Miguel, and Paco Calvo. 2023. "Plant Sentience? Between Romanticism and Denial: Science." *Animal Sentience* 33 (1). http://dx.doi.org/10.51291/2377-7478.1772.

Shaw, Robert E., and Jeffrey Kinsella-Shaw. 1988. "Ecological Mechanics: A Physical Geometry for Intentional Constraints." *Human Movement Science* 7 (2–4): 155–200. https://doi.org/10.1016/0167-9457(88)90011-5.

Shaw, Robert E., and Jeffrey Kinsella-Shaw. 2012. "Hints of Intelligence from First Principles." *Ecological Psychology* 24 (1): 60–93. https://doi.org/10.1080/10407413.2012.643725.

Shaw, Robert E., and Michael T. Turvey. 1981. "Coalitions as Models for Ecosystems: A Realist Perspective on Perceptual Organization." In *Perceptual Organization*, edited by Michael Kubovy, and James R. Pomerantz, 343–415: London: Routledge.

Stoffregen, Thomas A. 2003. "Affordances as Properties of the Animal-environment System. *Ecological Psychology* 15 (2): 115–34. https://doi.org/10.1207/S15326969ECO1502_2.

Strong, Donald R., and Thomas S. Ray. 1975. "Host Tree Location Behavior of a Tropical Vine (Monstera gigantea) by Skototropism." *Science* 190: 804–6. https://doi.org/10.1126/science.190.4216.804.

Swenson, Rod, and Michael T. Turvey. 1991. "Thermodynamic Reasons for Perception–Action Cycles." *Ecological Psychology* 3 (4): 317–48. https://doi .org/10.1207/s15326969eco0304_2.

Taiz, Lincoln, Daniel Alkon, Andreas Draghun, Angus Murphy, Michael Blatt, Chris Hawes, Gerhard Thiel, and David G. Robinson. 2019. "Plants Neither Possess nor Require Consciousness." *Trends in Plant Science* 24 (8): 677–87. https://doi.org/10.1016/j.tplants.2019.05.008.

Trewavas, Anthony. 2016. "Intelligence, Cognition, and Language of Green Plants." *Frontiers in Psychology* 7: 588.

Turvey, Michael T. 1992. "Affordances and Prospective Control: An Outline of the Ontology." *Ecological Psychology* 4 (3): 173–87.

Turvey, Michael T. 2018. *Lectures on Perception: An Ecological Perspective.* New York: Routledge.

Turvey, Michael T., and Claudia Carello. 2012. "On Intelligence from First Principles: Guidelines for Inquiry into the Hypothesis of Physical Intelligence (PI)." *Ecological Psychology* 24 (1): 3–32. https://doi.org/10.1080/10407413 .2012.645757.

Ullman, Shimon. 1980. "Against Direct Perception." *Behavioral and Brain Sciences* 3 (3): 373–81. https://psycnet.apa.org/doi/10.1017/S0140525X0000546X.

Warren, William H. 1984. "Perceiving Affordances: Visual Guidance of Stair Climbing." *Journal of Experimental Psychology: Human Perception and Performance* 10 (5): 683. https://doi.org/10.1037//0096-1523.10.5.683.

Webre, Daniel J., Peter M. Wolanin, and Jeffry B. Stock 2003. "Bacterial Chemotaxis." *Current Biology* 13 (2): R47–9. https://doi.org/10.1016/s0960 -9822(02)01424-0.

7 Ecological Plant Learning

Jonny Lee and Aditya Ponkshe

7.1 Introduction

Ecological psychology is a scientific framework that offers an alternative to much of mainstream cognitive psychology (e.g., Gibson 1979; Michaels and Carello 1981; Reed 1988). Pioneered by J.J. and E.J. Gibson, ecological psychology began as a theory of perception (Gibson 1966) and perceptual development (Gibson 1969). The ecological approach replaces psychological explanations that appeal to inferential processes over internal representations by a more-or-less secluded nervous system, with appeals to the detection of rich ecological information available to a whole, continuously acting organism, within its environment (e.g., Turvey and Carello 1981).

Despite aspirations of generalisability across perceptual modalities and biological taxa, J.J. Gibson (1979) explicitly resisted the idea that plants perceive in a sense analogous to humans and other animals (cf. Calvo 2022; Calvo and Lawrence 2022). It is with some irony then that with its dismissal of the neural or computational basis of intelligence and emphasis on the emergent behaviour of a coupled organism-environment system (e.g., Turvey and Carello 2012), ecological psychology has provided a lens through which to view (and defend) the possibility of intelligent behaviour in non-neural organisms, including plants (Carello, Symons, and Martín 2012; Gagliano 2015; Segundo-Ortin and Calvo 2019, 2022; Calvo 2016; Calvo, Raja, and Lee 2017; Abramson and Calvo 2018; Calvo and Lawrence 2022).

The ecological approach to plants has thus far prioritised explanations of plant perception. However, reflecting the potential for an ecological theory of psychology more generally, and an ecological approach to learning especially, ecological tools have begun to be applied to other areas of research into plant intelligence (e.g., Calvo 2022). Thus, in addition to introducing the ecological approach to plants more generally, this chapter surveys the prospects of applying an ecological theory of learning to plants.

We proceed as follows. Sections 7.2 and 7.3 provide a brief overview of the ecological framework before introducing the basic principles of

DOI: 10.4324/9781003393375-11

plant behaviour, and how an ecological approach has proved explanatory. Sections 7.4 and 7.5 turn to ecological learning and its potential to provide an illuminating window into learning in plants. Section 7.6 concludes with some considerations for future research into plant learning.

7.2 Ecological Psychology

Ecological psychology offers a scientific framework for the study of perception and other psychological phenomena, such as learning and memory (for an overview of topics covered by ecological psychology, see Blau and Wagman 2022). Rather than treating the organism and its environment as separate entities, ecological psychology approaches them as a unit or system. In doing so, it appeals to the manner in which objects and events lawfully structure energy distributions which can be encountered by an active creature in a rolling cycle of action and perception (e.g., Gibson 1979; Michaels and Carello 1981). Therefore, ecological psychology begins with the assumption that the correct unit of analysis in psychology is the organism-environment. This approach began in large part as a theory of perception—or perception-action, because perception and action are taken to be locked in a loop of mutual constraint (e.g., Warren 2006). Specifically, many of the elements of the theory originated in the study of human and other animal perception (although, as we shall, these may be extended to include the perceptive abilities of non-zoological taxa).

One way to conceive of ecological psychology is to contrast it with the view of perception it is often cast in opposition to, namely, a cognitivist or inferentialist account (epitomised, for instance, by a Helmholtzian approach to perception; e.g., Rock 1983; cf. Carello, Symons, and Martín 2012). According to ecological psychology, perception is not best explained in terms of sensory stimuli received by passive organs at a discrete moment, nor the associated thought that perception requires reconstructing the world via enriching, inferential processes that compensate for impoverished stimuli resulting in ambiguous, static images. Instead, perception is achieved by a whole organism that is continuously acting across time, and in doing so encountering rich ecological information in its environment (or niche) through its senses, where such senses are active systems connected to the whole body rather than passive channels (e.g., Gibson 1966; Glotzbach and Heft 1982). Perceiving organisms are thus acting organisms—creatures constantly exploring their environment and so allowing for the continual uncovering of rich information.

To understand what makes such information rich, we can consider the primary variables involved in perception. Cognitivist approaches take perception to begin with stimuli understood in terms of lower-order variables, that is, properties like wavelength (in the case of vision), or amplitude (in the case of audition). Perceivers are then tasked with computing higher-order information that is meaningful to their needs (such as a complete three-dimensional image of a tree) based on lower-order variables. By contrast, ecological psychology suggests that perception begins with the detection of higher-order variables. Higher-order variables refer to complex patterns of information that often involve combinations of lower-order variables and can provide information about more abstract properties of the world. Events change over time and space and, as ecological psychologists note, the invariant patterns in these changes provide information about the event. Part of the task of ecological psychology becomes the identification of the right higher-order variables detected by a perceptual system for guiding action.

An important detail in this emerging ecological picture is the claimed specificity of the structured energy in an organism's environment. Creatures perceive by encountering (often through movement) patterns in lawfully structured arrays of ambient energy that specify their surroundings. These structured energy distributions are specific to (hence, specify) certain states of affairs or environments. An ecological understanding of vision, for instance, begins by noting the heterogeneous and often highly nuanced structure of the ambient optic array, that is, the lawfully patterned light that fills a medium (such as air) due to the reflectance properties, shape, orientation and other features of objects (e.g., Glotzbach and Heft 1982). This ambient optic array is informative in the sense that the structured energy is specific to the environment and a point of observation. In other words, the information available to the organism is specific to its particular surroundings (for a strong version of this idea, see Turvey et al. 1981). A range of morphological, perceptual and other biological and physical properties of the creature (e.g., its size, shape, number and kinds of limbs) will determine (1) which points of observation it occupies, and (2) how these points of observation change over time (e.g., see Blau and Wagman 2022, chap. 5). Given that perception is explained in terms of the detection of information specific to a particular state of affairs or environment, without appealing to mediating, inferential processes, it is said to be direct. This contrasts with indirect theories of perception where some internal process of enrichment is required to compensate for impoverished (and thus ambiguous) stimuli.

In animal vision, for instance, the optic flow is specific to and so informative about the direction and rate of movement. Optic flow here refers to the global pattern of change in an image due to the relative motion between an organism and its environment, i.e., the apparent motion of objects in a visual image as one moves (Gibson 1950). Enrichment is not required because creatures do not need to elaborate ambiguous lower-order variables. Rather, perception requires detecting sufficiently unambiguous (lawful) patterns in structured energy. From such observations, ecological psychologists suggest that structured energy arrays inform a creature about its relationship with the environment, and in particular, what affordances are available.

Such affordances are the primary objects of perception, according to ecological psychology, made possible by the aforementioned detection of lawfully structured information (e.g., Richardson et al. 2008). These are opportunities for behaviour that emerge from the interaction between organisms and their environment—possibilities for actions that an environment grants (affords) an organism such as climbing, hiding or swimming. When we look at a hammer, say, we do not perceive the medium itself per se (light, in this instance) nor an object abstracted from its usability (an object extended in space with such-and-such dimensions). Instead, we perceive a graspable tool. As Gibson (1979) helpfully summarises: "The affordances of the environment are what it offers the animal, what it provides or furnishes, either for good or ill [...] It implies the complementarity of the animal and the environment" (p. 127). Thus, organisms primarily perceive possibilities for action and so too how their environment relates to themselves, in contrast to an action-independent world on which significance is subsequently imposed by a separable cognitive process.

Ecological psychology is a scientific framework and is intended to generate experimental paradigms, predictions and research heuristics (e.g., for an early, important empirical application of affordance theory, see Warren 1984). One significant development within the ecological tradition, pioneered by David Lee in the 1970s, was Tau theory and its subsequent applications (for an introduction and overview, see Lee et al. 2009). Take the following problem: how does an organism know when an approaching object will make contact, for example, when a child is tasked with catching a ball? Ecological psychology predicts that the information to solve the problem is available in the ambient energy array of the creature's environment, specifically the optic array.

Developing this idea, ecological psychologists explain that the distance between an observer and an object is available in the relative rate of expansion of the object's image—more exactly, how quickly the approaching object expands relative to its current degree of occlusion of the background (Michaels and Oudejans 1992). Tau (τ) is the mathematical variable which describes the time an organism will take to close the gap given the current rate of closure. The defining equation is $\tau(X, t) = X(t)/\dot{X}(t)$ where X is the current magnitude of the gap to be closed and $\dot{X}$ is the current rate of change of X. "Tau dot", i.e., the rate $\dot{\tau}(X, t)$, can be further used to avoid the collision. Essentially, rather than internally calculating when the ball will make contact, for example, by subconsciously dividing the estimated distance over the estimated velocity (which would be lower-order variables), given the current magnitude of the gap and its rate of change, a child can know when the ball will make the contact. For this to happen, the child merely needs to detect how quickly the approaching ball is expanding in their visual field. As an object approaches, the image it projects onto the retina grows larger. Consequently, the rate at which this image expands can then be used to detect the time-to-contact without knowing anything about distance or speed.

General tau theory provides a generic version of this idea, explaining perceptual/motor control with the same theoretical scheme (Lee et al. 2009). In essence, controlled movement (in general) can be explained in terms of closing motion gaps, reflecting gaps in current and goal states. If the retinal image is expanding at a constant rate, tau remains constant, and the time-to-contact can be directly perceived from the value of tau. Though the theory was initially used to explain collision avoidance and preparation for contact, it has subsequently been applied to movement and control more generally and inspired the search for other variables in the optic array that may be exploited by an organism. As we shall see in Section 7.3, tau theory has been used to explain not only the behaviour of humans and other animals but also plants.

7.3 An Ecological Approach to Plant Behaviour

Ecological psychology is typically taken to be generalisable: it is a theory of perception across taxa (e.g., it generalises beyond human perception) and different perceptual modalities (e.g., it generalises beyond vision). Nevertheless, as noted above, J.J. Gibson rejected the idea that plants perceive, and ecological psychology is still often presented as a theory of animal perception (but see Carello, Symons, and Martín 2012). For example, in discussing the ambition of ecological psychology to offer a form of explanation across sensory modalities, Blau and Wagman (2022) write, "the ecological perspective argues that treating the different perceptual systems of a given animal as requiring different explanations is just as bad as treating the perceptual systems of different kinds of animals as requiring different explanations" (p. 65). The authors thus highlight that ecological psychology provides a type of explanation across modalities. However, they imply ecological psychology is restricted to animalia. Such plant blindness (Wandersee and Schussler 1999, 2001) arises mainly out of our inability to appreciate plant behaviour.

Unlike animals, plants are sessile and rooted; they cannot walk, run, swim, or fly. As a result, plants appear stationary which makes it harder to think of them as behaving. Classical definitions of behaviour, being zoo-centric, generally assume movement via locomotion as a prerequisite (Tinbergen 1951; see further examples discussed in Levitis, Lidicker, and

Freund 2009). Since plants do not locomote, it is easy to consider what plants do as something other than behaviour. However, locomotion is not the only way to move in a physical space. Instead of locomotion, plants move by growing and developing. While growing, plants integrate biotic and abiotic parameters (refer to Table 1 in Calvo and Trewavas (2021) for the detailed list of external and internal signals sensed by plants) and develop new parts and/or expand the current ones, with the help of totipotent meristematic tissue which is divisible (Silvertown and Gordon 1989). Growth-based movement coupled with highly recursive, modular, and decentralised body structure enables plants to flexibly change their phenotype (locally and/or globally) in a coordinated manner. Consequently, growth-based movement and developmental phenotypic plasticity are integral to plant behaviour which are otherwise excluded from the zoo-centric definitions of behaviour (Trewavas 2009; Cvrčková, Žárský, and Markoš 2016).

Plant movement is not a new idea. In a pioneering series of books, Darwin describes at length the habits of plants, focusing on their idiosyncratic movements (Darwin 1875; Darwin and Darwin 1880). Plants move either directionally where environmental stimuli determine the growth trajectory (e.g., phototropism), and/or non-directionally without necessarily involving growth (e.g., nyctinasty: circadian rhythm based non-directional movements, and thigmonasty: non-directional movements to touch or vibrations). Be it a sensitive plant (*Mimosa pudica*), rapidly folding its leaves in response to touch, or a carnivorous Venus flytrap (*Dionaea muscipula*) closing its traps, we can more readily appreciate plant movements observable on a human time scale. However, fast plant movements are rare. Most plant movements are slow and go unnoticed by the human eye. A closer look reveals that tips of shoots, leaves, flowers and roots continuously oscillate and actively exhibit movement of nutation while growing— nutation being the bending movement of plant organs caused as a result of differential growth in different parts of organs. Our inability to appreciate the continuous movement of plant body parts in some way accounts for why plant actions are traditionally seen as passive and reactive in nature (Calvo 2022).

Plant actions cannot afford to be merely reflexive, especially given that growth-based movements are irreversible. Through continuous action of nutation, plants actively engage with the environment and sense the wide range of biotic and abiotic factors to guide their growth in an anticipatory and goal-directed manner; be it roots searching for nutrients belowground or shoots of climbers searching for support. For instance, a growing tendril of the common bean (*Phaseolus vulgaris*) continuously revolves in oval-shaped trajectories (Darwin used the term circumnuation to describe revolving circular nutation movements; see Darwin (1875); Darwin and

Darwin (1880)) and extends itself to grab the support. Since growing while circumnutating is an irreversible process and metabolically costly, climbing beans cannot afford to miss the target. To succeed, it needs to control the trajectory of circumnutation (Raja et al. 2020). A combination of endogenous (e.g., cell elongation and sedimentation process mediated via phytohormones and changes in turgor pressure) and exogenous (e.g., gravity) processes contribute to the persistence of circumnutation (Caré et al. 1998; Mugnai et al. 2015; Stolarz 2009). However, it is not clear how climbing beans control their approach towards the target while circumnutating. A compelling idea is that rather than bumping up onto the support randomly, the approach towards support is followed by ecologically guided control laws such as the General Tau theory (Calvo, Raja, and Lee 2017; Frazier et al. 2020).

As introduced in Section 7.2, tau theory was originally used to explain how organisms prospectively control their approach movement while closing any kind of gap, be it a plummeting Gannet diving underwater to catch the fish (Lee and Reddish 1981), a fly landing on the surface (Wagner 1982), or humans gazing (Grealy, Craig, and Lee 1999) or catching objects (van der Meer, van der Weel, and Lee 1994). The theory suggests that while "closing the gap", organisms rely on a high-order variable tau which explains the time to close the gap by taking into account the current rate of closure. According to orthodox ecological psychology, the temporal information specified by tau is sufficient enough to guide organisms towards the target. The theory was initially restricted to animals but has since been applied outside the animal kingdom to unicellular protists (Lee et al. 2009; Delafield-Butt et al. 2012). Preliminary work on the guidance of circumnutation of climbing beans (Calvo, Raja, and Lee 2017) suggests that in principle, there is no reason why the theory would not apply to plants. Ecologically speaking, while growing around in an environment with structured energy fields of different sorts ranging from chemical gradients produced by volatile and non-volatile chemicals to electromagnetic and vibratory haptic fields (Calvo and Trewavas 2020), the relation between revolving beans and the environment can generate relational information about the support which in turn would specify further action for the climbing bean and guide it to reach the support. Thus, climbers can potentially detect the climbable support in their vicinity by detecting perceptual information generated as a result of coupling between revolving motion and the climbable support's properties. Once detected, climbing beans can then use this perceptual information to endogenously guide their approaching manoeuvre prescribed by tau.

Ecologically guided anticipatory actions are not only restricted to aerial parts of the plant body but can also work underground (Calvo, Martín, and Symons 2014). Take the example of pea plant roots. When given a choice between static vs temporally dynamic nutrient regimes, rather than opting for a static resource that is rich in absolute terms, young pea plants grow their roots anticipatory based on the future availability of nutrients. They invest more resources in roots from the patches where nutrient levels are increasing (not static) even when they are poor in absolute terms than other patches (Novoplansky 2016). In another instance, roots of the Velvetleaf plant (*Abutilon theophrasti*) can sense whether the nutrient distribution is homogenous versus heterogeneous; they can then combine this information with the presence/absence of competitors, and accordingly, decide whether to avoid or share nutrients with competitors by growing or not growing roots in the areas occupied by competitors (Cahill et al. 2010). Plant behaviour, therefore, can only be understood in light of growth-dependent and/or independent movements and phenotypic plasticity. Continuously moving plant body parts constantly perceive a wide range of biotic and abiotic factors and ecologically guide plant actions in an anticipatory and goal-directed way in response to a combination of endogenous and exogenous processes.

Plants are thus capable of behaviours we might consider "intelligent". Though a complete discussion is beyond the scope of this chapter, it is worth acknowledging that such intelligent behaviours are realised, in part, by internal processes with similarities to animals, from the fundamental role of action potentials in electrical signalling (Fromm and Lautner 2007; Lee and Calvo 2023) to the regulatory role of the same basic hormones (Ramakrishna and Roshchina 2018). This has been the subject of investigation by the emerging field of "plant neurobiology". Curiously, both ecological psychology and plant neurobiology have proceeded to study plant intelligence, more-or-less, in the absence of a cognitivist framework. It is perhaps no surprise then that the two have been suggested by some as potential partners, where plant neurobiology reveals the underlying plant components that help realise whole plant-environment interaction. As Calvo (2016) would have it, "plant neurobiology" is gradually revealing relevant similarities between plants and animals, in which both use functionally similar internal signalling systems to coordinate behaviours with the environment:

> We can, therefore, approach plant behavior and neurobiology from the point of view of ecological psychology, and analyze the plant-environment system as a whole whose behavior emerges and self-organizes at a particular scale of interaction, the one mandated by ecology.
>
> (p. 21)

We add that this is possible because—though titled plant *neuro*biology— the significance of the field lies partly in the very discovery of biological, physiological and functional similarities between animals and plants despite the latter lacking neurons (e.g., Brenner et al. 2006). Ecological psychology and plant neurobiology both begin their study of intelligent behaviour free from assumptions about the necessity of neural hardware or internal cognitive representations, and so it is unsurprising that ecological psychology and plant neurobiology may make comfortable bedfellows.

Returning to the point at hand, we have prima facie reason to consider that plants do indeed detect information in the ambient energy array and use it to control behaviour, conforming to the fundamentals of the ecological framework. Pressure is placed, therefore, on J.J. Gibson's demarcation between behaving animals and non-behaving plants.

7.4 Ecological Learning

Ecological psychology is typically intended to offer insight beyond traditional problems of perception and into other domains of psychology. This is based on two overlapping assumptions: (1) the founding principles of ecological psychology, such as organism-environment mutuality, have implications across psychological abilities, and (2) perception-action is at the root of many or all psychological abilities. These assumptions are evident in ecological learning. Extending the principles of perception to learning, ecological psychology shifts explanation away from mental inferences over stored representations and towards a constantly unfolding relationship between an active organism and its environment.

The foundations of an ecological approach to learning began with the pioneering work of E.J. Gibson on perceptual learning or perceptual development (Gibson 1969). Once again, E.J. Gibson's approach rejects the idea that learning should be approached with the assumption that the organism faces an ambiguous environment which must be compensated for through processes of enrichment (Gibson and Gibson 1955). Instead, perceptual learning is rooted in an organism's ability to develop the capacity to detect specific information for affordances and align their intended behaviour to that information. In essence, perceptual learning is the process of better aligning available information and behaviour.

The concept of affordances as the primary objects of perception thus provides a starting point for an ecological approach to learning (e.g., Lobo, Heras-Escribano, and Travieso 2018). Much of learning can be cast in terms of the development of affordances; simply put, an organism learns to complete some task, or improves its performance on some task, via the acquisition or refinement of the ability to use available information in an energy array to perceive an affordance and control possible actions. Ecological learning thus involves an organism developing, improving or calibrating the ability to perceive its relationship with the environment at an ecological scale. Learning is continuous with perception to the extent it depends on the same detection and use of structured patterns in the organism's environment (e.g., Blau and Wagman 2022, 195).

This general idea of an ecological approach to learning has been developed under the auspice of direct learning. In Jacob and Michael's (2007) formulation, the theory of direct learning seeks to explain exactly how learning is possible without appealing to inferential processes or a loan of intelligence (e.g., p. 330). Whereas theories of enrichment or "adding to" (Michaels and Carello 1981) might explain the superior performance of an expert over a novice (say, in a perceptual discrimination task) in terms of the contribution of knowledge stored in memory, ecological theories suggest that learning reflects mere changes in the alignment of organism and environment, "for example, changes in which properties of ambient energy arrays perceptual systems respond to" (Jacobs and Michaels 2007, 322). As such, the improved performance of an expert over an amateur, say, "derives from the improved fits of experts to their environments, rather than from an increased complexity of computational and memorial processes" (ibid.). While direct perception emphasises the (non-inferential) detection of information that specifies an environment or state of affairs, direct learning emphasises the (non-inferential) fine-tuning of that detection, and its use, in a manner that improves performance (Calvo 2022).

The idea that learning involves the fine-tuning (Blau and Wagman 2022, 204) of perception-action abilities in relation to an organism's environment led to the identification of three kinds of direct learning (Jacob and Michael 2007; Jacobs, Silva, and Calvo 2009), or three components within the direct learning process: education of intention, education of attention, and calibration.

Education of intention roughly refers to the development or fine-tuning of an organism's ability to select between a range of actions (appropriate to the situation), and so in turn the selection of which affordances to attend to. Given the ecological theory of perception, this involves the active uncovering of information in the environment; determining to fend off a predator rather than run away entails the (rather hasty) exploration of the energy arrays to uncover the pertinent information for that particular action.

Education of attention roughly refers to the development or fine-tuning of an organism's ability to detect and attend to information in the patterns of an energy array that is more pertinent to the affordance in question. In the process of learning to bat a ball, one might improve their ability to detect information about the ball's time of arrival.

Calibration roughly refers to the development or fine-tuning of an organism's ability to map the detection of information in patterns in an energy array to appropriate perception-action. The concept of calibration reflects the fact that detecting stimulation patterns does not automatically imply the ability to use (or use well) those patterns for perceiving and acting. Education of intention and education of attention only results in adaptive developments of an organism's ability to perceive and act if there is an alignment between the uncovering and detecting of information in the environment and the organism's perceptual and motor capacities. Particularly throughout ontogenetic development, organisms are likely to learn to calibrate information in their environment with perception-action.

Calvo, Symons and Martín (2012) helpfully summarise by noting that, from within an ecological framework, intention "sets the goals and boundary constraints of actions, and hence defines which actions can be considered well adapted" whereas "attention, understood ecologically, refers to the informational basis of perception and action at a particular moment" (p. 2). Calibration is the third component that captures developing connections between the detection of increasingly specific information variables and the execution of relevant actions.

As we have seen, direct learning involves a change in the informational basis of perception and action via education of attention. Consequently, the ecological framework suggests that with practice, learners gradually move from less specific to more specific informational variables and rely on more useful variables as practice progresses (for related discussion, see Jacobs, Silva, and Calvo 2009). This change from less specific to more specific informational variables allows for performance improvements. Whether the change in the informational variables during practice occurs slowly or quickly depends on the specificity and usefulness of information variables. If a particular informational variable is highly specific and useful, then changes in the variable throughout practice will be slower whereas if the informational variable is relatively less specific and less useful, then the changes will be quicker. Huet et al. (2011) illustrate this hypothesis in the context of flight landing. They show that participants with no previous flight experience learn simulated landing manoeuvres quicker under variable than constant practice conditions, which they take to be due to education of attention. While explaining why variable practice conditions make practice more optimal than constant conditions, they argue that "Variability of practice reduces the usefulness of initially used informational variables, which leads to a quicker change in variable use, and hence to a larger improvement in performance" (p. 1841). A follow-up experiment in which experimental factors were selectively varied further led to the supporting claim that "Participants tended to converge toward the variables that were useful in the specific conditions that they encountered during practice" (ibid.). However, note that whether plants show similar behavioural patterns remains an unexplored question. We return to this below.

The theory of direct learning thus underscores the importance of advancing towards more specific informational variables during repeated practice of perceptual-motor tasks and explains how the nature of the informational landscape influences the rate of learning. Below, we will see how these lessons may cast light on learning in plants and help make sense of observations stretching back to Darwin.

7.5 Ecological (Plant) Learning

In casting learning as continuous with its characterisation of perception, ecological psychology provides prima facie grounds to consider learning in non-neural organisms. This is because the character of learning is not determined by, say, representational operations, or any animal-specific hardware, such as a nervous system (Carello, Symons, and Martín 2012). What matters, in a nutshell, is whether there is significant fine-tuning of behaviour via the detection of information in the ambient energy array. Ecological psychology thus invites us to imagine learning in non-neural organisms. This then raises the question of whether there is any evidence for plant learning.

Though now largely the purview of plant scientists, the earliest experiments into so-called plant learning were curiously conducted by comparative psychologists investigating the possibility of a generalised learning phenomena (cf. Abramson and Chicas-Mosier 2016). The two types of learning most studied in plants are habituation and classical conditioning, with the former receiving the greatest attention. Early studies of habituation experimented on *Mimosa pudica*, the sensitive plant which is remarkable for the speed with which it folds its leaves in response to mechanical stimulation. Early work by Pfeffer (1873) and Bose (1906) supported the claim that *Mimosa* exhibited decreased sensitivity following repeated stimulation, ruling out that desensitisation was the result of mere fatigue, and evidencing the ability of *Mimosa* to discriminate between types of stimuli (for discussion of earlier work on *Mimosa*, see Calvo 2022). Throughout the 20th century, Mimosa continued to function as an experimental organism for investigating habituation, given the apparent virtue of the species in overcoming the perennial obstacle of slow movement in studying plant behaviour (however, *Mimosa* presents its own methodological issues; see Abramson and Chicas-Mosier 2016, 3). Early classical conditioning was also investigated in Mimosa but with mixed results (e.g., compare the Haney study cited in Applewhite (1975), versus Levy, Caton, and Holmes (1970); see Abramson and Chicas-Mosier (2016), for discussion). A more recent and much-discussed paper by Gagliano et al. (2016) supplied evidence for associative learning in garden peas, *Pisum sativum*. The peas appeared to demonstrate a learned association between blue light (unconditioned stimulus) and airflow generated from a fan (conditioned stimulus). Though it has generated excitement, Markel (2020a) failed to replicate the study (but see Gagliano et al. 2020; Markel 2020b), and multiple methodological issues have been identified during the replication attempts, which are ongoing at the time of writing (Ponkshe et al. 2023).

Against this backdrop, what can an ecological approach contribute to the study of learning in plants? Speaking very generally, ecological psychology affects our guiding assumptions regarding what learning is and what things are capable of learning. First, as already suggested above, ecological psychology presents fewer conceptual barriers to entertaining the possibility of plant learning given its deemphasis on the neural basis of learning or the requirement of representational processes (Carello, Symons, and Martín 2012). For example, those with cognitivist inclinations might consider apparent evidence of associative learning in plants as not evidence of true learning because that requires underlying representational processes that are present in humans but absent in plants (Adams 2018). However, an ecological framework jettisons this assumption (as do certain behaviourist approaches), suggesting that even human learning may be non-representational in nature (Segundo-Ortin and Calvo 2019; Abramson and Calvo 2018). Second, ecological psychology underscores that at least much of learning involves the development of perceptual-motor abilities, organised around the perception of affordances, arguably evident in plants (see §4).

Beyond such sweeping considerations, how might ecological psychology contribute to the study of learning in plants? Calvo (2022) helpfully identifies two concrete proposals. First, he suggests, an ecological approach banishes the idea of equivalence of associability, following some contemporary behaviourists. According to this idea, associated with the classic studies of Pavlov, all stimuli are candidates for conditioning, stimulus intensity aside (for an overview and complications, see Coleman 2007). Ecological psychologists warn against hasty assumptions in the failure to demonstrate association given that, put simply, different stimuli are of greater or lesser (or no) relevance (see also Abramson and Calvo 2018). Following (Michaels and Carello 1981), failure to associate cues may demonstrate the absence of a related affordance. According to ecological psychology (as well as some versions of behaviourism), both the content of a cue and its context impact behavioural response. An obvious but still important lesson is the sensitivity to plant-relevant cues and contexts.

Second, Calvo (2022) indicates that if we assume ecological psychology provides the right framework with which to understand plant perception, then ecological psychology in turn helps to explain how plant learning is possible in the first place. This can be put in terms of a puzzle that applies to plants as much as animals: if perception is direct, unmediated and possible without inference, how might learning be similarly direct? Or put more strongly: if we assume intelligence generally operates without processes of mediation or enrichment, how is learning possible? Direct learning theory comports with the direct theory of perception and suggests that changes carried out during learning are specific to the environmental informational properties. We add that, by accessing task-specific information, learners tend to select invariants specific to previously selected ones over multiple trials. This proposition explains how performance improves with practice and how it can be attuned to the changes in the environment-learner unit. It is this aspect of direct learning, we suggest, that can be applied in the context of plant learning.

The idea is illustrated, we propose, by the phenomenon of climbing plants reaching for support. As surveyed in Section 7.3, climbing plants continuously circumnutate (continuously revolve in oval-shaped trajectories) while growing. As they grow, the pattern of circumnutation changes. This is not a newly reported phenomenon. Darwin himself had noticed continuous changes in nutation patterns. While describing the circumnutation patterns of Ceropegia, he noted:

> When a tall stick was placed so as to arrest the lower and rigid internodes of the Ceropegia, at the distance at first of 15 and then of 21 inches from the center of the revolution, the straight shoot slowly and gradually slid up the stick, so as to become more and more highly inclined, but did not pass over the summit. Then, after an interval sufficient to have allowed of a semi-revolution, the shoot suddenly bounded from the stick and fell over to the opposite side or point of the compass, and reassumed its previous slight inclination. It now recommenced revolving in its usual course, so that after a semi-revolution it again came into contact with the stick, again slid up it, and again bounded from it and fell over to the opposite side. This movement of the shoot had a very odd appearance, as if it were disgusted with its failure but was resolved to try again.
>
> (Darwin 1875, 12–13)

Calvo, Raja and Lee (2017) observed a similar swaying away movement pattern in common beans.

Darwin's early observations taken together with the recent monitoring of circumnutation patterns captured using sophisticated time-lapse photography suggest that climbing plants (e.g., the common bean) try again and again to reach the perceived support during multiple trials. We suggest that plants can select similar "target-specific" invariants over multiple trials and can improve their goal-directed movement performance with practice. It is in this process of refinement in performance over practice that there is room for speculating about the application of direct learning principles by which plants can be attuned to the changes in the "target environment-plant" unit. This is analogous to the participants in the landing manoeuvres learning experiments surveyed in the previous section. To test this idea, future work should focus on designing studies to show how differently climbing plants respond to different conditions with an emphasis on whether their performance is more optimal in variable than constant environmental conditions. Furthermore, studies should focus on identifying higher-order invariants that plants use to perceive the target in different environmental contexts.

In summary, though the study of plant behaviour and intelligence remains in its infancy, ecological psychology has the potential to inform our understanding of plant learning. It does this both by offering a general theoretical framework and by guiding experimental work.

7.6 Conclusion

The range and complexity of plant behaviour—especially that which may be considered "intelligent"—is of growing interest to researchers. Inevitably, debate over which theories allow us to best conceptualise and guide research on plant behaviour has followed. This chapter explored how the framework of ecological psychology has been applied to plants with a focus on perception, and how it might be extended to encompass plant learning.

With its rejection of certain cognitivist assumptions that problematise the notion of plant intelligence, the conceptual openness of ecological psychology to unorthodox forms of cognition provides prima facie support for its value in researching plant learning. We saw that concepts from direct learning theory, in particular education of intention, education of attention, and calibration, can help to make sense of learning in the absence of neural hardware or internal representations. More concretely, we indicated that existing experimental work on direct learning may be extended to investigate goal-directed growth in climbing plants. As with humans learning landing manoeuvres, we may be able to understand plants as becoming attuned to particular task-specific invariants with practice over time. The applicability of direct learning theory to climbing plant behaviour begs for further testing.

As empirical work on plant behaviour from an ecological perspective develops, theoretical work should also continue on its compatibility with plant neurobiology, given the rising prominence of the latter, and their accordant assumptions regarding the non-neural nature of intelligence. For instance, it could be shown how burgeoning work on plant electrical signaling ultimately enables communication across the whole plant, facilitating behaviour at the ecological scale. Given the emerging similarities between animal and plant physiology, this may even inform our understanding of how ecological explanations of animal behaviour connect with neurobiological explanations, and thus the relationship between ecological psychology and other sciences more broadly.

Acknowledgements

We are grateful to Paco Calvo for the discussion throughout the project and for providing insightful feedback. We would also like to thank anonymous reviewers for providing helpful comments and suggestions.

References

Abramson, Charles I., and Paco Calvo. 2018. "General Issues in the Cognitive Analysis of Plant Learning and Intelligence." In *Memory and Learning in Plants*. Signaling and Communication in Plants, edited by Frantisek Baluska, Monica Gagliano-, and Guenther Witzany, 35–49. Cham: Springer. https://doi .org/10.1007/978-3-319-75596-0_3
Abramson, Charles I., and Ana M. Chicas-Mosier. 2016. "Learning in Plants: Lessons from Mimosa pudica." *Frontiers in Psychology* 7. https://doi.org/10 .3389/fpsyg.2016.00417
Adams, Fred. 2018. "Cognition Wars." *Studies in History and Philosophy of Science Part A* 68: 20–30. https://doi.org/10.1016/j.shpsa.2017.11.007
Applewhite, Phillip B. 1975. "Learning in Bacteria, Fungi, and Plants." In *Invertebrate Learning. Cephalopods and Echinoderms*, vol. 3, edited by W. C. Corning, J. A. Dyal, and A. O. D. Willows, 179–86. New York and London: Plenum Press.
Blau, Julia J. C., and Jeffrey B. Wagman. 2022. *Introduction to Ecological Psychology: A Lawful Approach to Perceiving, Acting, and Cognizing*. New York: Routledge and CRC Press. https://doi.org/10.4324/9781003145691.
Bose, Jadish Chandra. 1906. *Plant Response*. New York and Bombay: Longmans, Green, and Co.
Brenner, Eric D., Rainer Stahlberg, Stefano Mancuso, Jorge Vivanco, František Baluška, F., Elizabeth Van Volkenburgh. 2006. "Plant Neurobiology: An Integrated View of Plant Signaling." *Trends in Plant Science* 11 (8): 413–19.
Cahill, James. F., Gordon G. McNickle, Joshua G. Haag, Eric G. Lamb, Samson M. Nyanumba, and Colleen Cassady St. Clair. 2010. "Plants Integrate Information About Nutrients and Neighbors." *Science*, 328 (5986): 1657–57. https://doi.org /10.1126/science.1189736
Calvo, Paco. 2016. "The Philosophy of Plant Neurobiology: A Manifesto." *Synthese* 193 (5): 1323–43. https://doi.org/10.1007/s11229-016-1040-1
Calvo, Paco. 2022. "Rompiendo con el zoocentrismo: hacia una psicología ecológica vegetal." In *Affordances y ciencia cognitiva: Introducción, teoría y aplicaciones*, edited by Manuel Heras Escribano, Lorena Lobo Navas, and Jesús Vega Encabo, 184–200. Tecnos.
Calvo, Paco, and Natalie Lawrence. 2022. *Planta Sapiens: Unmasking Plant Intelligence*. London: The Bridge Street Press.
Calvo, Paco, Emma Martín, and John Symons. 2014. "The Emergence of Systematicity in Minimally Cognitive Agents." In *The Architecture of Cognition: Rethinking Fodor and Pylyshyn's Systematicity Challenge*, edited by Paco Calvo and John Symons, 397–434. Cambridge, MA: MIT Press.
Calvo, Paco, Vicente Raja, and David N. Lee. 2017. "Guidance of Circumnutation of Climbing Bean Stems: An Ecological Exploration", *bioRxiv* 122358. https:// doi.org/10.1101/122358.

Calvo, Paco, John Symons, and Emma Martín. 2012. "Beyond 'Error Correction'." *Frontiers in Psychology* 3. https://doi.org/10.3389/fpsyg.2012.00423

Calvo, Paco, and Anthony Trewavas. 2020. "Physiology and the (Neuro)biology of Plant Behavior: A Farewell to Arms." *Trends in Plant Science* 25 (3): 214–16. https://doi.org/10.1016/j.tplants.2019.12.016.

Calvo, Paco, and Anthony Trewavas. 2021. "Cognition and Intelligence of Green Plants. Information for Animal Scientists." *Biochemical and and Biophysical Research Communications* 564: 78–58. https://doi.org/10.1016/j.bbrc.2020.07.139.

Caré, Anne-Frangoise, Leonid Nefed'ev, Bernard Bonnet, Bernard Millet, and Pierre-Marie Badot. 1998. "Cell Elongation and Revolving Movement in Phaseolus vulgaris L. Twining Shoots." *Plant and Cell Physiology* 39 (9): 914–21. https://doi.org/10.1093/oxfordjournals.pcp.a029454.

Carello, Claudia, Daniela Vaz, Julia J. C. Blau, and Stephanie Petrusz. 2012. "Unnerving Intelligence." *Ecological Psychology* 24 (3): 241–64. https://doi.org/10.1080/10407413.2012.702628

Coleman, S. R. 2007. "Pavlov and the Equivalence of Associability in Classical Conditioning." *The Journal of Mind and Behavior* 28 (2): 115–33.

Cvrčková, Fatima, Viktor Žárský, and Anton Markoš. 2016. "Plant Studies May Lead Us to Rethink the Concept of Behavior." *Frontiers in Psychology* 7. https://doi.org/10.3389/fpsyg.2016.00622

Darwin, Charles. 1875. *The Movements and Habits of Climbing Plants.*[2] London: John Murray.

Darwin, Charles, and Francis Darwin. 1880. *The Power of Movement in Plants.* London: John Murray.

Delafield-Butt, Jonathan T., Gert-Jan Pepping, Colin D. McCaig, and David N. Lee. 2012. "Prospective Guidance in a Free-Swimming Cell." *Biological Cybernetics* 106 (4): 283–93. https://doi.org/10.1007/s00422-012-0495-5.

Frazier, P. Adrian, Lorenzo Jamone, Kaspar Althoefer, and Paco Calvo. 2020. "Plant Bioinspired Ecological Robotics." *Frontiers in Robotics and AI* 7. https://doi.org/10.3389/frobt.2020.00079.

Fromm, Jörg, and Silke Lautner. 2007. "Electrical Signals and Their Physiological Significance in Plants." *Plant, Cell & Environment* 30 (3): 249–57.

Gagliano, Monica. 2015. "In a Green Frame of Mind: Perspectives on the Behavioural Ecology and Cognitive Nature of Plants." *AoB Plants* 7: plu075. https://doi.org/10.1093/aobpla/plu075.

Gagliano, M., V. Vladyslav Vyazovskiy, Alexander A. Borbély, Mavra Grimonprez, and Martial Depczynski. 2016. "Learning by Association in Plants." *Scientific Reports* 6 (1): 38427. https://doi.org/10.1038/srep38427.

Gagliano, Monica, Vladyslav Vyazovskiy, V., Alexander A. Borbély, Martial Depczynski, and Ben Radford. 2020. "Comment on 'Lack of Evidence for Associative Learning in Pea Plants'." *ELife* 9: e61141. https://doi.org/10.7554/eLife.61141.

Gibson, Eleanor. J. 1969. *Principles of Perceptual Learning and Development.* New York: Appleton-Century-Crofts.

Gibson, James. 1950. *The Perception of the Visual World.* Boston: Houghton-Mifflin.

Gibson, James. 1966. *The Senses Considered as Perceptual Systems.* Boston: Houghton-Mifflin.

Gibson, James J. 1979. *The Ecological Approach to Visual Perception* (pp. xiv, 332). Boston: Houghton-Mifflin.

Gibson, James J., and Eleanor J. Gibson. 1955. "Perceptual Learning: Differentiation or Enrichment?" *Psychological Review* 62: 32–41. https://psycnet.apa.org/doi/10.1037/h0048826.

Glotzbach, Phillip A., and Harry Heft. 1982. "Ecological and Phenomenological Contributions to the Psychology of Perception." *Noûs* 16 (1): 108–21. https://doi.org/10.2307/2215421.

Grealy, Madeleine A., Cathy M. Craig, and David N. Lee. 1999. "Evidence for On-line Visual Guidance During Saccadic Gaze Shifts." *Proceedings of the Royal Society of London. Series B: Biological Sciences* 266 (1430): 1799–804. https://doi.org/10.1098/rspb.1999.0849.

Huet, M., David M Jacobs, Cyril Camachon, Olivier Missenard, Rob Gray, and Gilles Montagne. 2011. "The Education of Attention as Explanation of Variability of Practice Effects: Learning the Final Approach Phase in a Flight Simulator." *Journal of Experimental Psychology: Human Perception and Performance* 37 (6): 1841–54. https://doi.org/10.1037/a0024386

Jacobs, David M., and Claire F. Michaels. 2007. "Direct Learning." *Ecological Psychology* 19 (4): 321–49. https://doi.org/10.1080/10407410701432337

Jacobs, David M., Paula L. Silva, and Juan Calvo. 2009. "An Empirical Illustration and Formalization of the Theory of Direct Learning: The Muscle-Based Perception of Kinetic Properties." *Ecological Psychology* 21 (3): 245–89. https://doi.org/10.1080/10407410903058302

Lee, David N., Reinoud J. Bootsma, Mike Land, David Regan, and Rob Gray. 2009. "Lee's 1976 Paper." *Perception* 38 (6): 837–58. https://doi.org/10.1068/pmklee

Lee, David N., and Paul E. Reddish. 1981. "Plummeting Gannets: A Paradigm of Ecological Optics." *Nature* 293: 293–4. https://doi.org/10.1038/293293a0

Lee, Jonny, and Paco Calvo. 2023. "The potential of plant action potentials." *Synthese*, 202 (6): 176.

Levitis, Daniel A., William Z. Lidicker, Jr., and Glenn Freund. 2009. "Behavioural Biologists Do not Agree on What Constitutes Behaviour." *Animal Behaviour* 78 (1): 103–10. https://doi.org/10.1016/j.anbehav.2009.03.018

Levy, E., A. Allen, W. Caton, and E. Holmes. 1970. "An Attempt to Condition the Sensitive Mimosa Pudica." *Biological Psychology* 12: 86.

Lobo, Lorenza, Manuel Heras-Escribano, and DavidTravieso. 2018. "The History and Philosophy of Ecological Psychology." *Frontiers in Psychology* 9. https://doi.org/10.3389/fpsyg.2018.02228

Markel, Kasey. 2020a. "Lack of Evidence for Associative Learning in Pea Plants." *ELife* 9: e57614. https://doi.org/10.7554/eLife.57614

Markel, Kasey. 2020b. "Response to Comment on 'Lack of Evidence for Associative Learning in Pea Plants'." *ELife* 9: e61689. https://doi.org/10.7554/eLife.61689

Michaels, Claire F., and Claudia Carello. 1981. *Direct Perception*. Englewood Cliffs, NJ: Prentice-Hall.

Michaels, Claire F., and Raoul R. D. Oudejans. 1992. "The Optics and Actions of Catching Fly Balls: Zeroing Out Optical Acceleration." *Ecological Psychology* 4 (4): 199–222. https://doi.org/10.1207/s15326969eco0404_1

Mugnai, Sergio, Elisa Azzarello, Elisa Masi, Camilla Pandolfi, and Stefano Mancuso. 2015. "Nutation in Plants." In *Rhythms in Plants: Dynamic Responses in a Dynamic Environment*, edited by Stefano Mancuso, and Sergey Shabala, 19–34. Cham: Springer. https://doi.org/10.1007/978-3-319-20517-5_2

Novoplansky, Ariel. 2016. "Future Perception in Plants." In *Anticipation Across Disciplines*, edited by Mihai Nadin, vol. 29, 57–70. Cham: Springer. https://doi.org/10.1007/978-3-319-22599-9_5

Pfeffer, Wilhelm. 1873. *Physiologische Untersuchungen*. Leipzig: W. Engelmann.

Ponkshe, Aditya, Jacobo Blancas Barroso, Charles I. Abramson, and Paco Calvo. 2023. "A Case Study of Learning in Plants: Lessons Learned from Pea Plants."

Quarterly Journal of Experimental Psychology. https://doi.org/10.1177/17470218231203078

Raja, Vicente, Paula L. Silva, Roghaieh Holghoomi, and Paco Calvo. 2020. "The Dynamics of Plant Nutation." *Scientific Reports* 10 (1): 19465. https://doi.org/10.1038/s41598-020-76588-z

Ramakrishna, Akula, and Victoria Vladimirovna Roshchina, eds. 2018. *Neurotransmitters in Plants: Perspectives and Applications.* Boca Raton: CRC Press.

Reed, Edward S. 1988. *James J. Gibson and the Psychology of Perception.* Yale University Press. https://doi.org/10.2307/j.ctt1xp3nmm

Richardson, Michael J., Kevin Shockley, Brett R. Fajen, Michael A. Riley, and Michael T. Turvey. 2008. "9 - Ecological Psychology: Six Principles for an Embodied–Embedded Approach to Behavior." In *Handbook of Cognitive Science,* edited by Paco Calvo and Antoni Gomila, 159–87. Elsevier. https://doi.org/10.1016/B978-0-08-046616-3.00009-8

Rock, Irvin. 1983. *The Logic of Perception.* Cambridge: MIT Press.

Segundo-Ortin, Miguel, and Paco Calvo. 2019. "Are Plants Cognitive? A Reply to Adams." *Studies in History and Philosophy of Science Part A* 73: 64–71. https://doi.org/10.1016/j.shpsa.2018.12.001

Segundo-Ortin, Miguel, and Paco Calvo. 2022. "Consciousness and Cognition in Plants." *Wiley Interdisciplinary Reviews: Cognitive Science* 13 (2): e1578. https://doi.org/10.1002/wcs.1578

Silvertown, Jonathan, and Deborah M. Gordon. 1989. "A Framework for Plant Behavior." *Annual Review of Ecology and Systematics* 20 (1): 349–66. https://doi.org/10.1146/annurev.es.20.110189.002025

Stolarz, Maria. 2009. "Circumnutation as a Visible Plant Action and Reaction." *Plant Signaling & Behavior* 4 (5): 380–7. https://doi.org/10.4161/psb.4.5.8293

Tinbergen, Nikolaas. 1951. *The Study of Instinct.* Oxford: Clarendon Press.

Trewavas, Anthony. 2009. "What Is Plant Behaviour?." *Plant, Cell & Environment* 32 (6): 606–16. https://doi.org/10.1111/j.1365-3040.2009.01929.x

Turvey, M. T., and Claudia Carello. 1981. "Cognition: The View from Ecological Realism." *Cognition* 10 (1–3): 313–21. https://doi.org/10.1016/0010-0277(81)90063-9

Turvey, M. T., and Claudia Carello. 2012. "On Intelligence From First Principles: Guidelines for Inquiry Into the Hypothesis of Physical Intelligence (PI)." *Ecological Psychology* 24 (1): 3–32. https://doi.org/10.1080/10407413.2012.645757

Turvey, M. T., Robert E. Shaw, E. S. Reed, and William M. Mace. 1981. "Ecological Laws of Perceiving and Acting: In Reply to Fodor and Pylyshyn (1981)." *Cognition* 9 (3): 237–304. https://doi.org/10.1016/0010-0277(81)90002-0

van der Meer, Audrey L. H., F. R. Ruud van der Weel, and David N. Lee. 1994. "Prospective Control in Catching by Infants." *Perception* 23 (3): 287–302. https://doi.org/10.1068/p230287

Wandersee, James H., and Elisabeth E. Schussler. 1999. "Preventing Plant Blindness." *The American Biology Teacher* 61 (2): 82–6. https://doi.org/10.2307/4450624

Wandersee, James H., and Elisabeth E. Schussler. 2001. "Toward a Theory of Plant Blindness." *Plant Science Bulletin* 47 (1): 2–9.

Wagner, Hermann. 1982. "Flow-field Variables Trigger Landing in Flies." *Nature* 297 (5862): 147–8. https://doi.org/10.1038/297147a0

Warren, William H. 1984. "Perceiving Affordances: Visual Guidance of Stair Climbing." *Journal of Experimental Psychology: Human Perception and Performance* 10 (5): 683–703. https://doi.org/10.1037/0096-1523.10.5.683
Warren, William H. 2006. "The Dynamics of Perception and Action." *Psychological Review* 113 (2): 358–89. https://doi.org/10.1037/0033-295X.113.2.358

8 On Plant Affordances

Gabriele Ferretti

Il n'y a pas d'autre mouvement en eux que l'extension. Aucun geste, aucune pensée, peut-être aucun désir, aucune intention, qui n'aboutisse à un monstrueux accroissement de leur corps, à une irrémédiable *excroissance*. Francis Ponge, *Faune et Flore*, in *Le parti pris des choses*

8.1 Introduction

Philosophy of mind and cognitive science are interested in understanding how biological systems register what is found in the external environment so as to successfully guide their motor behavior in order to survive. Many debates have grown on the investigation about how *accurate* the cognitive processing of living beings has to be in order to grant survival (Godfrey-Smith 1996; Dennett 1987), on the biological strategies deployed by cognitive mechanisms in different contexts to process information (Millikan 1984; Dretske 1981), and on the epistemological significance of the evolutionary dimension of these mechanisms (Angelucci et al. 2021, 2023; Hoffman, Singh, and Prakash 2015; Hoffman and Manish 2012). In this respect, the last few decades in the study of cognition have seen a revolution. The cognitivist idea that cognition is a functional system whose processing is not constituted by its material substrate has been replaced by the idea that cognition constitutively depends on the morphology of the body of the cognitive system, i.e., its embodiment, as well as on its biological evolution and its embedment in a specific environment (Newen, De Bruin, and Gallagher 2020; Shapiro 2019), or Um-welt (Berthoz and Christen 2008). These pieces of research permit to consider an evolutionary perspective on embodiment and cognition (Barrett 2018; Keijzer 2017; Wilson 2008; Marshall, Houser, and Weiss 2021).

An important role has been played by the idea that cognition is massively shaped by the deep interplay between perception and action granted by the embodiment of organisms (Hurley 1998; Chemero 2009; Noë 2004; Ferretti and Zipoli Caiani 2021a, 2021b, 2023; Engel, Friston, and Kragic 2016). Within this story, not by chance, the ecological theory of

DOI: 10.4324/9781003393375-12

perception by J.J. Gibson (1979/1986) is usually advocated to explain how animal perception is massively bound to action, as one of the main functions of perception, if not the main one, is that of successful action guidance. In the Gibsonian jargon, perception is about *affordances*, i.e., possibilities of action offered by objects in the environment that an animal can detect. As action is possible in relation to motoric skills given by bodily configuration, then the notion of affordance suggests the crucial link between *embodiment, action* and *perception*. This notion is precious in the light of the above-discussed idea that perception-action coupling may be at the basis of cognitive processes, in both humans and animals. Different varieties of sensori-motor behavior often are the target of those research programs investigating affordances.

There is also another revolution that has taken place in the last years. The debate on the embodied nature of cognition has attracted many scholars working on different types of biological organisms, capable of displaying some sort of minimal cognitive ability. This led to suggest, contrary to the original idea that cognition is an affair of neural systems hosted by human and non-human bodies, that also forms of non-neural cognition can exist (Calvo and Baluška 2015; Baluška and Levin 2016). Indeed, non-neural systems seem capable of performing cognitive processes that were usually ascribed only to neural systems with different (and more complex) bodily structures.

This revolution has also extended to the study of plants, hosting the interesting idea of a possibility for plants to exhibit cognition. Several scholars have indeed advocated the capacity for plants to display cognitive processes that we usually expect only from animals (Trewavas 2005, 2014; Calvo and Keijzer 2011; Calvo 2007).

A specific claim in this field of research is that plants can rely on a perception-action coupling, at the basis of cognition, very similar to the one used by animals (Calvo 2016; Carello et al. 2012). And, if true, this claim leads to speculate that plants should be capable of perceiving affordances (Ibid.).

In what follows, I first discuss some crucial aspects of affordance perception in the literature on human and non-human animals, whose consideration is an important starting point for those interested in the study of plant affordances (§8.2). I then mention some studies on plants that may be representative for the claim that plants can perceive affordances (§8.3). Finally, I discuss the most important questions future research on plants may want to investigate when analyzing the claim that plants can perceive affordances (§8.4).

8.2 Affordances

Affordances have been at the center of a profound discussion at the crossroad between philosophy, psychology, and cognitive and motor neuroscience (for different angles, see Borghi and Riggio 2015; Osiurak et al. 2017; de Wit et al. 2017; Sakreida et al. 2016; Chemero 2009; Ferretti 2019, 2021a; Zipoli Caiani and Ferretti 2017).

On the one hand, affordances have a relational, dispositional nature. They emerge from the encounter of a specific subject and a specific object. A given, peculiar object will display a specific possibility for action only to a given, specific agent, based on the features of both, which can meet in a given interaction. This can give rise to a motor event on the behavioral side of the agent, which is related to the possibility offered within that specific coupling. And, in turn, as the affordance is a perceived possibility for action, the agent can spot it based on its perceptual capacities, and on its bodily morphology and motor skills. For example, in front of a mug, a human may perceive the possibility of using a *grasping* action with the hand to manipulate the handle of the mug, especially with a *precision grip*, i.e., a grip performed with thumb and index finger.

In this respect, a portion of the literature suggested, following the Gibsonian idea, that affordances are not represented, as they are the results of a coupling of the organism with the environment in which the former perceives salient ecological aspects of the latter without any representational mediation (Chemero 2009; Heras-Escribano 2019).

On the other hand, the neuroscience of affordance suggested that, in order to perceive an affordance, it is required for the agent to have at her disposal an *information-processing* system devoted to this task. In this view, the information pertaining to the action possibility is *detected* by the visual system, and *processed* to grant that a proper motor act is generated with respect to the features of the object the cognitive system is dealing with. This view suggests, within some accounts, that affordances are *represented* in the brain.

Of course, there can be cases in which affordances are also mis-perceived or, in the representationalist jargon, mis-represented. This can be due, as we shall see, to different problems, from an impairment of the visuomotor system, to a mental disorder.

Let us take a look at these different dimensions of affordances. This will serve as a basis to offer some specific questions for the research on plant affordances in the foreseeable future.

8.2.1 Behavioral Dispositionality

One of the preliminary *dispositionalist* interpretations of affordances has been offered by Turvey (1992; see also Turvey et al. 1981). According to this view, affordances, *qua* possibilities for action, can be complemented by the embodied, perceptual-motor capacities of specific animals. Thus, an affordance is a sort of disposition emerging from the encounter and interaction between the animal and an aspect of the environment it inhabits. More precisely, a target in the environment affords a given motor action to an animal if, and only if, given some basic environmental circumstances, a specific property of the target (e.g., a spatial configuration) is *complemented* by a particular property of the animal (e.g., a bodily architecture that permits to act on that spatial configuration). This makes possible, for the animal's motor skills, to satisfy, with an appropriate motor behavior, the possibility of action offered by the target. The output generated by the bodily action of the agent that makes it capable of acting upon the target is called *effectivity*, with the proviso that the affordance is *actualized*.

Recalling the previous example about the human grasping the mug, the affordance is *actualized* by being *complemented* by the human sensorimotor skill, this leading to the *effectivity* of the grasping action, so that the mug can be grasped. The *graspability* of the mug relates to the *affordance* perceived, while the (overt) *act of grasping* refers to the *effectivity*, that is *actualized* by the motor skill of the agent, when the mug is *grasped*.

According to this view, given the correct environmental circumstances, so that the animal has the capacity to perceive and perform the action on the aspect of the target that affords such an action, and the object displays this aspect, the motor act recalled from the perceived affordance *necessarily actualizes* (for different and critical angles on this account, see Zipoli Caiani 2013; Michaels and Carello 1981; Chemero 2009; Shaw, Turvey, and Mace 1982; Stoffregen 2003). The notion of *necessity* here is not *modal*, but related to a *nomological* aspect, concerning the ecological laws of animal-object interaction. Given suitable ecological conditions, the animal will perceive and motorically respond to the action afforded by a given target (Ibid.).

This account has been criticized and replaced from another account, by Scarantino (2003), which does not consider affordances as followed by *necessary actualizations* emerging from the encounter of an animal and a suitable target for action, but rather as *probabilities of manifestations*. The *actualization* of affordances, in this view, *does not always occur* once an animal has the suitable motor resource to interact with an aspect of the environment that displays all the requested characteristics to be acted upon, which are indeed perceived by the animal. The *actualization* of affordances is just *likely to occur*.

These two accounts propose different angles on the way affordances are brought into being. According to the former, once an animal and a target meet, so that the motor capacities of the animal can suitably complement the properties of the target, perceived by the animal, then, this *always* gives rise to an affordance to be satisfied. According to the latter, once an animal and a target meet, so that the motor capacities of the animal can suitably complement the properties of the target, perceived by the animal, then this *may* give rise to an affordance to be satisfied.

I need to point out that there has been a huge debate on the specific theoretical commitments endorsed by these two positions, especially concerning the notions of *necessity* and *probability* when it comes to the *actualization* of an affordance. Here, I am offering a very general description, not taking into account all the technical disagreements, about this debate, proposed in the literature (for a review, see Heft 2001; Chemero 2009; Heras-Escribano 2019).

8.2.2 Perceptual-Motor Information-Processing

There are, as we have seen, dispositional dynamics of the animal-environment affordance relation. However, the animal needs to perceive the affordance. Some scholars, as said, suggested that, for the animal to perceive an affordance, some *information-processing* is needed. Current research in human and non-human animal neuroscience shows that, indeed, to detect action possibilities in the environment, sensory systems must be able to *process* spatial information with the purpose of using it to reliably guide action. This *information-processing* of spatial aspects of a target translated into motor commands comes in the form of (*visuomotor*) representations. The brain *represents* objects' spatial features *as* affordances, to motorically build the actions to satisfy them (Jacob and Jeannerod 2003; Borghi and Riggio 2015; Sakreida et al. 2016; Osiurak et al. 2017; Ferretti 2021a).

The use of representations is not perfectly in tune with the more ecological stance of the Gibsonian theory of affordances, which refuses any notion of information-processing in the form of an internal representation performed by the cognitive system, stating that affordances are directly perceived (Chemero 2009; Heft 2001; Heras-Escribano 2019).

However, whether or not this story on affordances can be told in a unified framework including both the *dispositional* and the *information-processing* accounts, as a matter of fact, the capability of properly registering the source of information coming from the external environment is a prerequisite for any biological, neural, cognitive system. And so is, especially, for the dimension of the cognitive system that is responsible for the encoding of affordances, so as to respond with appropriate actions (Sakreida et al. 2016; de Wit et al. 2017; Osiurak et al. 2017; Costall and Morris 2015; Ferretti 2021a; Borghi and Riggio 2015; Jacob and Jeannerod 2003).

This is supported by the discovery of specific networks, in the brain of humans and other animals, which are devoted to specific perceptual-motor information-processing, whose main function is that of *affordance extraction* from objects. These networks allow the animal's cognitive system to process visual information of an object and its properties and trigger the motor program (*visuomotor transformation*) that can be used to interact with the properties of the object, on the basis of the animal motor capacities and embodied morphology. Thus, the visuomotor brain can suitably transform the visual input into a motor output by *reading* the motoric richness of a given spatial arrangement (Jeannerod 2006; Sakreida et al. 2016; de Wit et al. 2017; Osiurak et al. 2017; Costall and Morris 2015; Ferretti 2016a, 2016b, 2016c, 2021a; Borghi and Riggio 2015; Jacob and Jeannerod 2003; Chong and Proctor 2020; Zipoli Caiani 2013, Ferretti and Zipoli Caiani, forthcoming). These visuomotor responses to affordances are in play when the object is located within the *peripersonal space* of action of the observer (Ibid.).

There may also be competition between different affordances (*affordance competition*) from the same target source (Cisek 2007; Cisek and Kalaska 2010; Ferretti 2016a). Looking at a mug, several action possibilities can be recalled. A precision grip on the handle. A power grip on the cylinder. Information-processing devoted to *affordance extraction* must be capable of selecting one of the many affordances with respect to the task the animal has to fulfill for its survival.

This, in turn, relates to the fact that the action possibility is recalled not only with respect to the general motor capacities of the subject, *per se*, but also in relation to the *semantic significance* of the object, with respect to the action goal (Zipoli Caiani and Ferretti 2017; Chong and Proctor 2020; Sakreida et al. 2016; de Wit et al. 2017). Grasping a pen to write, or grasping it to move it along the table, are different motor acts, related to different motor goals (with respect to different functional uses, or semantic properties), and the former of which takes into account the function of the pen.

8.2.3 Problems with Affordances

As happens with perception and action, both *affordance detection* and *affordance satisfaction* are not always successful. There are different ways in which the animal cannot correctly process (i.e., perceive, or act upon) the affordance. Here is a (non-exhaustive) list of how this can happen, so that *affordance perception becomes problematic*.

(1). *Illusion or Mis-Perception.* Detection of (illusory) affordances may take place in case there is none. Think about the case of *trompe l'oeil* illusions, where, for example, you have a depicted library on a wall that looks like a library in the flesh, affording action (Ferretti 2016a, 2016b, 2016c, 2020a, 2020b, 2021b, 2023). You have the impression the object offers interaction. But this is just an illusory impression.

(2). *(Neuro-)Functional Impairment.* Neurological disruptions of the visuomotor system may endanger the proper detection of affordances, or the interaction with objects displaying them. Since affordances have both a visual and a motor component, from the point of view of the information-processing detecting them (which is indeed taken as visuo-motor processing, where the visual component and the motor component work in interplay), this disruption may happen with respect to each of these two components displaying a problem. For example, this can happen in case visual recognition is working, but the motor information-processing linking the visual stimulus to a proper motor command is disrupted. In this case, one may be visually aware of the spatial qualities of objects which commonly lead to an affordance, but, nonetheless, not capable of satisfying it with a proper motor act. Or this can also happen when motor processing is not impaired, but vision is not properly registering the scene and/or the target, as in the case of different forms of blindness (for discussion of these cases, see Jacob and Jeannerod 2003; Ferretti 2017, 2021c). In this case, one may still have intact the motor capacities to move, in principle, in such a way as to perform the motor act requested by the spatial arrangement of the object, but cannot correctly visually process the object, as well as the spatial information commonly associated with a motor performance upon it.

(3). *Pathological Affordance Responses.* Mental disorders may also endanger a proper relation with affordances, leading to different pathologies concerning affordance perception. In the case of *obsessive-compulsive-disorder*, affordances are restricted only to the compulsions afflicting the subject, while the others remain in the background, so that there is a pathological selection, based, for example, on fear and anxiety, of those action possibilities among the many others effectively offered from the visual scene. In contrast, in depression, the plethora of calls to action one can experience are lowered (de Haan et al. 2015, 2013). Moreover, *schizophrenic* patients display an impairment in visuomotor simulation (Sevos et al. 2013). Finally, in the case of *utilization behavior*, subjects feel forced to give rise to action performance recalled by an object. The way of interacting is proper, but the context of the performance is not appropriate. It is simply impossible to resist to the call for action of an affordance they spot. They feel they wanted to perform the action, even when this is not requested (for an analysis of this pathology, also in relation to other similar syndromes, see Eslinger 2002; Archibald, Mateer, and Kerns 2001; Ishihara et al. 2002; Iaccarino, Chieffi, and Iavarone 2014; Price and Libon 2011, Ferretti and Zipoli Caiani, forthcoming).

(4). *Broken Affordances.* There may also be *broken affordances*, where the objects recalling an affordance are damaged, and so they evoke *aversive* or *dangerous* action possibilities, the subject being aware that, if the action is satisfied, she may hurt herself during object manipulation.

In some cases, the spatial aspects that may hurt the subject are paid attention to, but not selected for action, while visual processing is oriented to the spatial properties of the object that can be manipulated (a broken glass may still exhibit an intact, i.e., non-damaged portion, where manipulability is possible). Here the damaged part offers a way of switching the motor act selected, with respect to the semantics (cfr. above). In other cases, the damage is so uniformly distributed that no action is possible, and the potential risks conflict with any alleged potential action recalled. There is no safe way of interacting with the object (Buccino et al. 2009; Borghi and Riggio 2015; Anelli, Borghi, and Nicoletti 2012; Ferretti and Chinellato 2019; Chinellato, Ferretti, and Irving 2019).

Taken together, these cases suggest the many problems potentially standing in the way of successful *affordance detection* and *satisfaction*, as well as the many constraints to be respected in order to obtain affordance perception for suitable interaction. These cases have been discussed at the crossroad of an interdisciplinary debate on the notion of affordance, which may be impossible to summarize in the present chapter, and which concerns optical, neurophysiological, computational, psychological, behavioral, phenomenological and pathological aspects of affordances (Osiurak et al. 2017; de Wit et al. 2017; Chong and Proctor 2020; Ferretti 2021a; Jacob and Jeannerod 2003, Ferretti and Zipoli Caiani, forthcoming).

However, what I've been saying can be precious for the researcher interested in plants, who wants to address the possibility of affordance perception in their case. Indeed, up to now, I've offered a sort of current *geography* of (the many aspects of) *affordance* perception. But the chapter wants to analyze *new pathways* in the world of plant affordances. For this reason, in the next section, I'll offer a brief overview of the studies concerning sensori-motor behavior in plants, which may be informative for the idea that they can perceive affordances. This will permit, in the last section of the chapter, to list a series of crucial questions, emerging from the coupling between the constraints on affordances listed so far and the evidence on plants I am about to discuss. This plethora of questions will constitute, indeed, a *theoretical map* to orient those future scholars who will pursue the interest in investigating whether plants perceive affordances.

8.3 Sensori-motor Behavior in Plants

Perceiving affordances, we've seen, depends as much on the animal sensory system as it depends on its bodily configuration and motor skills. A big portion of the experimental literature on human sensori-motor behavior, at the basis of affordances, is on grasping (Castiello 2005; Chinellato and del Pobil 2016). This literature parallels the one on grasping affordances in monkeys (Orban and Caruana 2014). Of course, examples of affordances are not exhausted with this motor act (Rietveld and Kiverstein 2014). But most of the research from visual and motor science has been made on grasping affordances. Indeed, we have a very elegant description of the information-processing in the human and monkey brain for grasping affordances (Castiello 2005; Ferretti 2021a; Sakreida et al. 2016; Borghi and Riggio 2015; Chinellato and del Pobil 2016; Orban and Caruana 2014). Then, this represents an important case study for those interested in philosophical reflections on affordances (Ferretti 2021a).

What about plants? Are there instances of sensori-motor behavior which may be related to affordance perception? Can there be a similar, paradigmatic example in their case, as the one of grasping for humans? Let's go slow on this.

As affordances depend on perception and action, it is first important to say something on these components in plants. First of all, plants have a sensory dimension. They can be sensitive to different sources of stimulation from the environmental context they are in (light, gravity, temperature) (Calvo 2016; Chamovitz 2012; Stolarz 2009), so as to display a form of perception (Chamovitz 2012; Karban 2015), which sometimes also recalls visual perception (Gavelis et al. 2015; Baluska and Mancuso 2016; Hayakawa et al. 2015) – of course, the move from sensory processing to perception (especially visual perception) may be a bit controversial in the light of philosophy of perception.

Interestingly, they also display a motor behavior permitting to actively respond to such stimuli (Ibid.). However, unlike animals, (most) plants don't change, properly speaking, the position of their "body" in space by moving away from a spatial location to another, toward a target, or away from a predator. They don't run or walk away to approach a source of needs, with respect to a given stimulus. They grow toward or away from different targets, for different reasons (Calvo 2016; Carello et al. 2012; Trewavas 2005). Plants can also generate simple forms of movements with their "body", using different organs, from leaves to roots (Mancuso and Shabala 2007; Gerbode et al. 2012; Stolarz 2009), in different contexts, for different environmental purposes, related to specific stimuli. Moreover, plants can also display movements, or *nutations*, that are not triggered by external stimuli, and are indeed performed autonomously. Among these, a famous one is that of *circumnutation*, a helical movement of the plant, due to the *elongation* of the organs, which can be performed in different scenarios (Stolarz 2009).

These are all clear examples of plant movements (sometimes related to sensory information). The interesting question, at this point, is whether there is any motor action, performed by plants, that is allegedly related to a response to an affordance, which is perceived. This would represent an interesting case study to consider here. Of particular relevance would be, moreover, to find an example, in plants, similar to the case of grasping in human affordances.

Of course, when we talk about grasping actions, an interesting and parallel case, which does not concern growing, is that of carnivorous plants (Böhm et al. 2016), as they can track and grasp their preys, with the purpose of eating them, thanks to mechanosensory responses of their inner surfaces stimulated by the insect touching them.

Recently, however, a similar interest on grasping in humans and monkeys has been devoted to the phenomenon of plant climbing (Calvo et al. 2014; Yokawa and Baluška 2018; Calvo and Friston 2017; Isnard and Silk 2009; Gerbode et al. 2012; Guerra et al. 2019; Raja et al. 2020; Scher, Holbrook, and Silk 2001; Saito 2022; Calvo, Raja, and Lee 2017) as a paradigmatic case of motor behavior that can be connected to affordances (Calvo 2016). Interestingly, in this respect, climbing and grasping can be related in the case of plants, and so can be their alleged affordances. Let's see how.

One of the most important pieces of evidence for research on plant affordances, in the case of climbing, which pantomimes grasping, comes from a study by Guerra et al. (2019), which shows that, in some cases of plants exhibiting climbing skills, the tendrils (whose form is like a two-digit appendage, cfr. p. 1) manage the structure of the support they climb upon. Indeed, the kinematics of movement concerning the aperture of the tendrils can be organized, during circumnutation, with respect to the dimensions, in particular the thickness, of the support. This would suggest there is a strict link between perceptually gained information and motor commands shaped upon it, so that the movement is controlled until the performance fully unfolds.

Interestingly, this study investigates climbing in plants in a very similar way with respect to which grasping is usually studied in humans. The researchers aim at verifying whether plants rely on actions that are driven by the intrinsic properties of the target. The study of the kinematics of movement of the tendrils' aperture in spatial proximity of the support parallels the study of the act of the grip aperture of the hand in reaching for grasping. Indeed, the initial aperture of the appendage, with the consequent closure as the target is completely matched, can be studied in the same manner as the behavior of the hand is. In this respect, there seems to be a relation between the features of the target and the biomechanics of "grasping" in both cases (Guerra et al. 2019, 1, 2).

One of the scenarios advocated by the authors may be that of a "visuo-motor transformation in which the visual coding of the object's intrinsic properties (e.g., thickness) is transformed into a pattern of movement" (p. 4). In this case, the evidence would tell us of sensory (*visual-like*) mechanisms performed by plants that can make use of spatial information to guide the motor action during climbing. (Let me point out that this hypothesis is highly speculative, though very fascinating.)

Another relevant study, in a related research path, comes from Saito (2022), which showed that, in climbing plants, tendrils display flexibility in coiling with respect to the diameter of the support, and this facilitates their motor behavior. This is crucially important for climbing plants, as the author suggests, since there is a correlation between the success of the climbing action and the diameter of the support upon which the plant grows, in relation to the length of the tendrils. The study, indeed, suggests that the tendrils coil not with respect to shape or size but with respect to the diameter (cfr. p. 6). This diameter-based specificity grants proper grip when the plant contacts the support. This also suggests that the plant behavior is capable of motorically dealing with the protrusions of the support, as to perform a suitable motor performance. This is possible thanks to "the preparation for phase into a clip-shaped gripping pattern" (p. 7). As the author notes:

> To discuss the advantages of clip shape coiling, it is necessary to consider that the natural supporting structures are unlike the smooth cylindrical supports used in the experiment, and they usually have many protrusions on their surface. It is easy to imagine that a small circle generated by the initial coiling (clip tip) acts like a hook and gets caught in the protrusion of the structure.
>
> (p. 7)

This also suggests a sort of decision making, in plants, for motor purposes, and on the basis of spatial information related to the support. This plant movement has several characteristics related to grasping, especially concerning grip aspects.[1]

Finally, another interesting experimental result for this discussion comes from Raja et al. (2020), where the authors show how climbing activity is shaped by the proximity of a support. This is highly interesting in the light of the fact that, as said, the processing of a gasping response, in humans and monkeys, depends on the target being located in the subject's peripersonal action space.

These experimental results seem to suggest that plants can engage in sensory-guided motor behavior. This may open to the possibility that they can detect affordances. This seems to be also suggested by the case of what we may define as a sort of *climbing-by-grasping* in plants, which is guided by some sort of minimal sensory processing of the properties of the support to guide movements.

So far so good. Now, I've first listed some important constraints on, and some crucial aspects of affordance perception. Then, I've discussed some relevant pieces of empirical evidence on plants, which may be crucial when investigating whether plants can perceive affordances, as these results flag some sensori-motor behavior in plants. In the next section, I propose a list of the questions we must consider when pursuing this investigation, which are formulated by bearing in mind the theoretical inventory on the many aspects of affordances offered in the first part of the chapter. Tackling these questions, in the light of the evidence on sensori-motor behavior just presented, can help us to understand whether there is reliable room for the claim that plants can perceive affordances.

8.4 On Plants and Affordances

Being plants capable of relying on sensory information for movement purposes (even if, most of the time, by growing), it may be tempting to claim that they can perceive, in some manner, affordances. The case of climbing seems to be a particularly informative one, as a parallel to grasping in humans. Thus, we may be inclined to suggest, in the case of plants, that an empty space affords growing, a support affords climbing, the body of a prey affords biting. Without any doubt, these interesting cases of plant sensori-motor behavior offer several precious hints, which may fuel the debate in the field of research on affordances. But investigating the possibility of affordances in plants must consider the many facets of the notion of affordance. In what follows, I want to flag, in the light of the considerations on the complex nature of affordances offered above (§8.2), some crucial questions that may orient the scholar interested in the claim that plants perceive affordances. A claim that may take many different theoretical routes, as many as the aspects of affordances are.

First off, with respect to dispositional accounts (§8.2.1), which dispositional theory would fit a description of plant behavior? Suppose there is a support that is detected from the plant, which may interact with it. How is the behavior shaped? Will a climbing act necessary follow? Or will the action only potentially actualize? In this case, can we account for the probabilities of manifestation of the action?

Second, in relation to information-processing (§8.2.2), what kind of information-processing do plants rely on when they guide their motor behavior towards the possibilities of interaction they may perceive, offered by a specific target in the environment? Can it be comparable to the one used by animals? Is it actually reasonable to talk about a similar form of *perceptual-motor transformation* generated by means of a chemical process? We saw that also plants can modulate the responses with respect to supports found in the proximity. What is the specific relation of these responses with respect to an alleged plant *"peripersonal" action space*?

Does the motor response correlate with "peripersonal" stimuli as in the case of animals? Consider also the case in which there are many supports, with multiple affordances. What is the best description of the plant behavior in this case? How does the selection of a given affordance work? Can there be an *affordance competition* with respect to a given scenario? And is the selection of the action carried with respect to different *semantics* of action? After all, a support may be used in different manners, for different purposes, as a pencil can, in case of humans.

Third, in the light of the problems that can afflict affordances (§8.2.3), are there cases of illusory affordances, with spatial stimuli leading the plant to process (and mis-perceive) an action possibility where there is none? (Cfr. point 1, §8.2.3) Are there cases where the plant cannot perceive an affordance due to damage or impairment in the (alleged) information-processing for affordances? (Cfr. point 2, §8.2.3) Is there any case of pathological affordance behavior? (Cfr. point 3, §8.2.3) And, concerning aversive/broken affordances, are there specific behaviors in which the plant detects a stimulus potentially related to an action, but whose performance may damage the plant's body? What does the plant do in this case? (Cfr. point 4, §8.2.3)

Finally, and more generally, to what extent is the coupling between the plant and a given support actually a response to an affordance? Action requires an agent. Is this just a basic agential response? Can we have a theory of action in plants capable of explaining their agential status, even if related to a form of a minimal motor behavior?

Of course, another interesting question is about whether plants require representations of the external environment to perceive it and guide their action upon it. It would be then also interesting to frame the question about affordance perception, realized by plants, in the context of the debate on representations. In this respect, the investigation on whether *all* animal affordances are detected or not with representational strategies is matter of controversy, especially in the light of the literature in cognitive science (Sakreida et al. 2016; Jacob and Jeannerod 2003; de Wit et al. 2017; Costall and Morris 2015; Ferretti 2021a; Borghi and Riggio 2015; Osiurak, Rossetti, and Badets 2017, Ferretti and Zipoli Caiani, forthcoming). If we can consider, within the literature on affordances, also plants and other non-neural biological systems, then, things are even more interesting, and we may need another story on perception and action. An intuitive way of describing these plant phenomena would be, of course, that of embracing a minimal view on cognition, as with many forms of non-neural systems, suggesting that plant sensori-motor behaviors concern a perfect case of environment-organism system where no representational task is requested (Calvo 2016). On the other hand, we should not exclude that, if plants rely on an information-processing, this may be realized by means of representations (maybe at the chemical level). Only future philosophical treatment can disclose this theoretical box on plant cognition.

To conclude, the debate on plant affordances may become a natural, theoretical research path these new case studies lead us to. The aim of this chapter was that of offering conceptual and theoretical tools to the researcher interested in affordances in plants. The idea was that of (i) considering the enormous complexity of the notion of affordance, and (ii) mentioning those results about plants that may potentially represent an interesting case of affordance perception, with the proviso that (iii) we should be capable of answering very specific and different questions on affordances in plants, based on what we know about affordances and their different aspects in animals. That said, it is not excluded that the mechanisms responsible for plant affordances may be very different from the mechanisms involved in animal affordances, and that the study on plants may tell us something new about the way in which affordance perception can take place.

8.5 Conclusion

This chapter offers some basic insights for any potential reasoning on the phenomenon of plant affordances. These insights stem from what we know from the traditional literature on affordances, concerning their many facets and, most important, the constraints they exhibit. Building a coherent theory of plant affordances requires explicitly mentioning these constraints, and then showing whether and how plants can meet them. In particular, the literature must be capable of explaining which behavioral dispositions plants have toward targets offering potential action, which kind of information-processing guides these motoric dispositions, and whether this can sometimes misfunction, or be broken. This seems to be the challenge that must be successfully met by any potential theory of plant affordances.

The chapter does not propose any specific claim on plant affordances, in this respect. The aim here is just to flag possible research paths to be pursued, with respect to the many directions of the literature on affordances, for the experts in the field of plants. I hope researchers on plant behavior will be interested in taking time to consider (at least some of) these questions, as to explicitly work on the notion of plant affordance, with respect to its possible manifestations, and in the light of the numerous empirical results on plant perception and action.

Acknowledgments

I want to thank Peter Schulte and Paco Calvo, whose comments allowed me to improve the chapter. I also want to thank the audience of the conference

Green Intelligence, Debating Plant Cognition, at the University of Basel. This work was supported by a German Humboldt Fellowship, hosted by Professor Albert Newen at the Institute for Philosophy II, Ruhr-University Bochum, Germany.

Note

1 This relates to grasping also in the light of the fact that the study suggests "the possibility of creating a new plant-type robot arm that can grasp objects of various shapes using an autonomous decentralised unit" (p. 7). This is interesting because robots that pantomime plants' motor behavior have been built (Lee and Calvo 2021), and also human grasping has been replicated, with robotic arms, in an embodied manner (Chinellato and Del Pobil 2016; Ferretti and Chinellato 2019; Chinellato, Ferretti, and Irving 2019).

References

Anelli, Filomena, Anna M. Borghi, and Roberto Nicoletti. 2012. "Grasping the Pain: Motor Resonance with Dangerous Affordances." *Consciousness and Cognition* 21 (4): 1627–39. https://doi.org/10.1016/j.concog.2012.09.001.

Angelucci, A., Vincenzo Fano, Gabriele Ferretti, Roberto Macrelli and Gino Tarozzi. 2023. "Does Evolution Favor Accurate Perception?" [Special Issue], *Argumenta* 9 (1): 105–14. https://doi.org/10.14275/2465-2334/202317.ang.

Angelucci, Adriano, Vincenzo Fano, Gabriele Ferretti, Roberto Macrelli, and Gino Tarozzi. 2021. "Evolutionary Dynamics and Accurate Perception. Critical Realism as an Empirically Testable Hypothesis." *Philosophia Scientiae* 25: 157–78. https://doi.org/10.4000/philosophiascientiae.2960.

Archibald, Sarah. J., Catherine A. Mateer, and Kimberley A. Kerns. 2001. "Utilization Behavior: Clinical Manifestations and Neurological Mechanisms." *Neuropsychology Review* 11 (3): 117–30. https://doi.org/10.1023/a:1016673807158.

Baluška, František, and Michael Levin. 2016. "On Having no Head: Cognition Throughout Biological Systems." *Frontiers in Psychology* 7: 902. https://doi.org/10.3389/fpsyg.2016.00902.

Baluška, František, and Stefano Mancuso. 2016. "Vision in Plants Via Plant-specific Ocelli?" *Trends in Plant Science* 21: 727–30. https://doi.org/10.1016/j.tplants.2016.07.008.

Barrett, Louise. 2018. "The Evolution of Cognition." In *The Oxford Handbook of 4E Cognition*, edited by Albert Newen, Leon De Bruin, and Shaun Gallagher, 710–34. Oxford Library of Psychology (online edn, Oxford Academic, October 9, 2018). https://doi.org/10.1093/oxfordhb/9780198735410.013.38

Berthoz, Alain, and Yves Christen. 2008. *Neurobiology of "Umwelt. How Living Beings Perceive the World*. Berlin, Heidelberg: Springer https://doi.org/10.1007/978-3-540-85897-3.

Böhm, Jennifer, Sönke Scherzer, Elzbieta Krol, Ines Kreuzer, Katharina von Meyer, Christian Lorey, and Thomas D. Mueller et al. 2016. "The Venus Flytrap Dionaea muscipula Counts Prey- Induced Action Potentials to Induce Sodium Uptake." *Current Biology* 26: 286–95. http://dx.doi.org/10.1016/j.cub.2015.11.057.

Borghi, Anna M., and Lucia Riggio. 2015. "Stable and Variable Affordances Are Both Automatic and Flexible." *Frontiers in Human Neuroscience* 9. https://doi.org/10.3389/fnhum.2015.00351.

Buccino, Giovanni, Marc Sato, Luigi Cattaneo, Francesca Rodà, and Lucia Riggio. 2009. "Broken Affordances, Broken Objects: A TMS Study." *Neuropsychologia* 47 (14): 3074–8. https://doi.org/10.1016/j.neuropsychologia.2009.07.003.

Calvo, Paco. 2007. "The Quest for Cognition in Plant Neurobiology." *Plant Signaling & Behavior* 2 (4): 208–11. https://doi.org/10.4161/psb.2.4.4470.

Calvo, Paco. 2016. "The Philosophy of Plant Neurobiology: A Manifesto." *Synthese* 193: 1323–43. https://doi.org/10.1007/s11229-016-1040-1.

Calvo, Paco, and František Baluška. 2015. "Conditions for Minimal Intelligence Across Eukaryota: A Cognitive Science Perspective." *Frontiers in Psychology* 6: 1329. https://doi.org/10.3389/fpsyg.2015.01329

Calvo Paco, and Karl Friston. 2017. "Predicting Green: Really Radical (Plant) Predictive Processing." *Journal of the Royal Society Interface* 14 (131): 20170096. https://doi.org/10.1098/rsif.2017.0096

Calvo, Paco, and Fred Keijzer 2011. "Plants: Adaptive Behavior, Root-brains, and Minimal Cognition." *Adaptive Behavior* 19 (3): 155–71. https://psycnet.apa.org/doi/10.1177/1059712311409446.

Calvo, Paco, Emma Martín, and John Symons. 2014. "The Emergence of Systematicity in Minimally Cognitive Agents." In *Systematicity and Cognitive Architecture: Conceptual and Empirical Issues 25 Years after Fodor & Pylyshyn's Challenge to Connectionism*, edited by Paco Calvo, and John Symons, 397–433. Cambridge, MA: MIT Press.

Calvo, Paco, Vicente Raja, and David N. Lee. 2017. "Guidance of Circumnutation of Climbing Bean Stems: An Ecological Exploration." *bioRxiv* 122358. https://doi.org/10.1101/122358.

Carello, Claudia, Daniela Vaz, Julia J. C. Blau, and Stephanie Petrusz. 2012. "Unnerving Intelligence." *Ecological Psychology* 24 (3): 241–64. https://doi.org/10.1080/10407413.2012.702628.

Castiello, Umberto. 2005. "The Neuroscience of Grasping." *Nature Reviews* 6 (9): 726–36. http://dx.doi.org/10.1038/nrn1744.

Chamovitz, Daniel. 2012. *What a Plant Knows: A Field Guide to the Senses*. New York: Scientific American Books.

Chemero, Anthony P. 2009. *Radical Embodied Cognitive Science*. Cambridge, MA: MIT Press.

Chinellato, Eris, and Angel P. del Pobil. 2016. *The Visual Neuroscience of Robotic Grasping. Achieving Sensorimotor Skills through Dorsal-ventral Stream Integration*. Cham: Springer.

Chinellato Eris, Gabriele Ferretti, and Lucy Irving. 2019. "Affective Visuomotor Interaction: A Functional Model for Socially Competent Robot Grasping." In *Biomimetic and Biohybrid Systems. Living Machines 2019. Lecture Notes in Computer Science 11556*, edited by Uriel Martinez-Hernandez, Vasiliki Vouloutsi, Anna Mura, Michael Mangan, Minoru Asada, Tony J. Prescott, and Paul F. M. J. Verschure, 51–62. Cham: Springer.

Chong, Isis, and Robert W. Proctor. 2020. "On the Evolution of a Radical Concept: Affordances According to Gibson and Their Subsequent Use and Development." *Perspectives on Psychological Science* 15 (1): 117–32. https://doi.org/10.1177/1745691619868207.

Cisek, Paul. 2007. "Cortical Mechanisms of Action Selection: The Affordance Competition Hypothesis." *Philosophical Transactions of the Royal Society, Biological Sciences* 362: 1585–99. http://dx.doi.org/10.1098/rstb.2007.2054.

Cisek, Paul, and John F. Kalaska. 2010. "Neural Mechanisms for Interacting with a World Full of Action Choices." *Annual Review of Neuroscience* 33: 269–98. https://doi.org/10.1146/annurev.neuro.051508.135409.

Costall, Alan, and Paul Morris. 2015. "The 'Textbook Gibson: The Assimilation of Dissidence."*History of Psychology* 18 (1): 1–14. https://doi.org/10.1037/a0038398.

de Haan Sanneke, Erik Rietveld, Martin Stokhof, and Damiaan Denys. 2015. "Effects of Deep Brain Stimulation on the Lived Experience of Obsessive-Compulsive Disorder Patients: In-Depth Interviews with 18 Patients." *PLoS One* 10 (8): e0135524. doi:10.1371/journal. pone.0135524.

de Haan, Sanneke, Erik Rietveld, Martin Stokhof and Damiaan Denys (2013). The phenomenology of deep brain stimulation-induced changes in OCD: an enactive affordance-based model. Frontiers in Human Neuroscience. doi: 10.3389/fnhum.2013.00653.

de Wit, Matthieu M., Simon de Vries, John van der Kamp, and Rob Withagen. 2017. "Affordances and Neuroscience: Steps Towards a Successful Marriage." *Neuroscience and Biobehavioral Reviews* 80: 622–9. https://doi.org/10.1016/j.neubiorev.2017.07.008

Dennett, Daniel C. 1987. *The Intentional Stance*. Cambridge, MA: MIT Press.

Dretske, Fred. 1981.*Knowledge and the Flow of Information*, Cambridge, MA: MIT Press.

Engel, Andeas K., Karl J. Friston, and Danica Kragic, eds. 2016. *The Pragmatic Turn. Toward Action-Oriented Views in Cognitive Science*. Cambridge, MA and London: MIT Press.

Eslinger Paul J. 2002. "The Anatomic Basis of Utilisation Behaviour: A Shift from Frontal-parietal to Intra-frontal Mechanisms." *Cortex* 38 (3): 273–6. https://doi.org/10.1016/S0010-9452(08)70658-0.

Ferretti, Gabriele. 2016a. "Through the Forest of Motor Representations." *Consciousness and Cognition* 43: 177–96. https://doi.org/10.1016/j.concog.2016.05.013.

Ferretti, Gabriele. 2016b. "Visual Feeling of Presence." *Pacific Philosophical Quarterly* 99 (7–8): 112–36. http://dx.doi.org/10.1111/papq.12170.

Ferretti, Gabriele. 2016c. "Pictures, Action Properties and Motor Related Effects." *Synthese, Special Issue: Neuroscience and Its Philosophy* 193 (12): 3787–817. https://link.springer.com/article/10.1007/s11229-016-1097-x

Ferretti, Gabriele. 2017. "Two Visual Systems in Molyneux Subjects." *Phenomenology and the Cognitive Sciences* 17 (4): 643–79. https://doi.org/10.1007/s11097-017-9533-z.

Ferretti, Gabriele. 2019. "Visual Phenomenology versus Visuomotor Imagery: How Can we be Aware of Action Properties?" *Synthese* 198: 3309–38. https://doi.org/10.1007/s11229-019-02282-x.

Ferretti, Gabriele. 2020a. "Why *Trompe l'oeils* deceive our Visual Experience." *The Journal of Aesthetics and Art Criticism* 78 (1): 33–42. https://doi.org/10.1111/jaac.12688.

Ferretti, Gabriele. 2020b. "Do *Trompe l'oeils* Look Right When Viewed from the Wrong Place?" *The Journal of Aesthetics and Art Criticism* 78 (3): 319–30. https://doi.org/10.1111/jaac.12750.

Ferretti, Gabriele. 2021a. "A Distinction Concerning Vision-for-Action and Affordance Perception." *Consciousness and Cognition* 87: 103028. https://doi.org/10.1016/j.concog.2020.103028.

Ferretti, Gabriele. 2021b. "Why the Pictorial needs the Motoric." *Erkenntnis* 88: 771–805. https://doi.org/10.1007/s10670-021-00381-1.

Ferretti, Gabriele. 2021c. "On the Content of Peripersonal Visual Experience." *Phenomenology and the Cognitive Sciences* 21: 487–513 https://doi.org/10.1007/s11097-021-09733-2.

Ferretti, Gabriele. 2023. "For an Epistemology of Stereopsis." *Review of Philosophy and Psychology*. https://doi.org/10.1007/s13164-023-00711-y.

Ferretti, Gabriele, and Eris Chinellato. 2019. "Can Our Robots Rely on an Emotionally Charged Vision-for-Action? An Embodied Model for Neurorobotics." In *Blended Cognition, The Robotic Challenge*, edited by Jordi Vallverdú and Vincent C. Müller, 99–126. Springer Series in Cognitive and Neural Systems, vol. 12. Cham: Springer.

Ferretti, Gabriele, and Silvana Zipoli Caiani, eds. 2021a. "Between Vision and Action. Special Issue." *Synthese* 98 (17). https://link.springer.com/journal/11229/volumes-and-issues/198-17/supplement.

Ferretti, Gabriele, and Silvano Zipoli Caiani. 2021b. "How Knowing-That and Knowing-How Interface in Action: The Intelligence of Motor Representations." *Erkenntnis* 88: 1103–33. https://doi.org/10.1007/s10670-021-00395-9.

Ferretti, Gabriele, and Silvano Zipoli Caiani. 2023. "The Rationality and Flexibility of Motor Representations in Skilled Performance." *Philosophia* 51: 2517–42. https://doi.org/10.1007/s11406-023-00693-2.

Ferretti, Gabriele and Silvano Zipoli Caiani, (Forthcoming). An All-Purpose Framework for Affordances. Reconciling the Behavioral and the Neuroscientific Stories. *Synthese, Special Issue: Neuroscience and Its Philosophy*.

Gavelis, Gregory S., Shiho Hayakawa, Richard A. White III, Takashi Gojobori, Curtis A. Suttle, Patrick J. Keeling, and Brian S. Leander. 2015. "Eye-like Ocelloids Are Built from Different Endosymbiotically Acquired Components." *Nature* 523: 204–7. https://doi.org/10.1038/nature14593.

Gerbode, Sharon J., Joshua R. Puzey, Andrew G. McCormick, and L. Mahadevan. 2012. "How the Cucumber Tendril Coils and Overwinds." *Science* 337 (6098): 1087–91. https://doi.org/10.1126/science.1223304.

Gibson, James. 1979/1986. *The Ecological Approach to Visual Perception*. Hillsdale: Lawrence Erlbaum Associates.

Godfrey-Smith, Peter. 1996. *Complexity and the Function of Mind in Nature*. Cambridge: Cambridge University Press.

Guerra, Silvi, Peressotti, Alessandro Peressotti, Francesca Peressotti, Maria Bulgheroni, Walter Baccinelli, Enrico D'Amico, Alejandra Gómez, Stefano Massaccesi, Francesco Ceccarini, and Umberto Castiello. 2019. "Flexible Control of Movement in Plants." *Scientific Reports* 9: 16570. https://doi.org/10.1038/s41598-019-53118-0.

Hayakawa, Shiho, Yasuharu Takaku, Jung Shan Hwang, Takeo Horiguchi, Hiroshi Suga, Walter Gehring, Kazuho Ikeo, and Takashi Gojobori. 2015. "Function and Evolutionary Origin of Unicellular Camera-type Eye Structure." *PLoS One* 10 (3): e0118415. https://doi.org/10.1371/journal.pone.0118415.

Heft, Harry. (2001). Ecological Psychology in Context. James Gibson, Roger Barker, and the Legacy of William James's Radical Empiricism. New York: Routledge.

Heras-Escribano, Manuel. 2019. *The Philosophy of Affordances*. Palgrave Macmillan.

Hoffman, Donald D., and Manish Singh. 2012. "Computational Evolutionary Perception." *Perception* 41 (9): 1073–91. https://doi.org/10.1068/p7275.

Hoffman, Donald D., and Manish Singh, Chetan Prakash. 2015, "The Interface Theory of Perception." *Psychonomic Bulletin & Review* 22 6: 1480–506. https://doi.org/10.3758/s13423-015-0890-8.

Hurley, Susan L. 1998. *Consciousness in Action*. Cambridge, MA: Harvard University Press.

Iaccarino, Leonardo, Sergio Chieffi, and Alessandro Iavarone. 2014. "Utilization Behavior: What Is Known and What Has to Be Known?" *Behavioural Neurology* 2014: e297128. https://doi.org/10.1155/2014/297128.

Ishihara, Kenji, Hiroshi Nishino, Toshiyuki Maki, Mitsuru Kawamura, Shigeo Murayama. (2002). "Utilization Behavior as a White Matter Disconnection Syndrome." *Cortex* 38 (3): 379–87. https://doi.org/10.1016/s0010 -9452(08)70666-x.

Isnard, Sandrine, and Wendy K. Silk. 2009. "Moving with Climbing Plants from Carles Darwin's Time into the 21st Century." *American Journal of Botany* 96 (7): 1205–21. https://doi.org/10.3732/ajb.0900045.

Jacob, Pierre, and Marc Jeannerod. 2003. *Ways of Seeing. The Scope and Limits of Visual Cognition.* Oxford: Oxford University Press.

Jeannerod, Marc. 2006. *Motor Cognition: What Actions Tell the Self.* Oxford: Oxford University Press.

Karban, Richard. 2015. *Plant Sensing and Communication.* Chicago: The University of Chicago Press.

Keijzer, Fred A. 2017. "Evolutionary Convergence and Biologically Embodied Cognition." *Interface Focus* 7 (3): 20160123. https://doi.org/10.1098/rsfs .2016.0123.

Lee, Jonny, and Paco Calvo. 2021. "Enacting Plant-Inspired Robotics." *Frontier in Neurorobotics* 15: 772012. https://doi.org/10.3389/fnbot.2021.772012.

Mancuso Stefano, and Sergey Shabala. 2007. *Rhythms in Plants: Phenomenology, Mechanisms and Adaptative Significance.* Berlin, Heidelberg: Springer.

Marshall, Peter J., Troy M. Houser, and Staci M. Weiss. 2021. "The Shared Origins of Embodiment and Development." *Frontiers in Systems Neuroscience* 15. https://www.frontiersin.org/article/10.3389/fnsys.2021.726403.

Michaels, Claire F., and Claudia Carello. 1981. *Direct Perception.* Englewood Cliffs, NJ: Prentice-Hall.

Millikan, Ruth G. 1984. *Language, Thought and Other Biological Categories.* Cambridge, MA: MIT Press.

Newen, Albert, Leon De Bruin, and Shaun Gallagher, eds. 2020. *The Oxford Handbook of 4E Cognition.* Oxford: Oxford University Press.

Noë, Alva. 2004. *Action in Perception.* Cambridge, MA: MIT Press.

Orban, Guy A., and Fausto Caruana. 2014. "The Neural Basis of Human Tool Use." *Frontiers in Psychology* 5 (310). http://dx.doi.org/10.3389/fpsyg.2014 .00310.

Osiurak, François, Yves Rossetti, and Arnaud Badets. 2017. "What is an Affordance? 40 Years Later." *Neuroscience, Biobehavioral Reviews* 77: 403– 17. https://doi.org/10.1016/j.neubiorev.2017.04.014.

Price Catherine, C., and David J. Libon. 2011. "Utilization Behavior." In *Encyclopedia of Clinical Neuropsychology,* by Jeffrey S. Kreutzer, John DeLuca, and Bruce Caplan, New York, NY: Springer, p. 2579.

Raja, Vicente, Paula L. Silva, Roghaieh Holghoomi, and Paco Calvo. 2020. "The Dynamics of Plant Nutation." *Scientific Reports* 10: 19465. https://doi.org/10 .1038/s41598-020-76588-z.

Rietveld, Erik, and Julian Kiverstein. 2014. "A Rich Landscape of Affordances." *Ecological Psychology* 26 (4): 325–2. https://doi.org/10.1080/10407413.2014 .958035.

Saito, Kazuya. 2022. "A Study on Diameter-dependent Support Selection of the Tendrils of *Cayratia japonica.*" *Scientific Reports* 12: 4461. https://doi.org/10 .1038/s41598-022-08314-w.

Sakreida, Katrin, Isabel Effnert, Serge Thill, Mareike M. Menz, Doreen Jirak, Claudia R. Eickhoff, Tom Ziemke, Simon B. Eickhoff, Anna M. Borghi, and Ferdinand Binkofski. 2016. "Affordance Processing in Segregated Parieto-frontal

Dorsal Stream Sub-Pathways." *Neuroscience and Biobehavioral Reviews* 69: 89–112. https://doi.org/10.1016/j.neubiorev.2016.07.032.

Scarantino, Andrea. 2003. "Affordances Explained." *Philosophy of Science* 70 (5): 949–61. https://doi.org/10.1086/377380.

Scher, Julia L., N. Michele Holbrook, and Wendy K. Silk. 2001. "Temporal and Spatial Patterns of Twining Force and Lignification in Stems of Ipomoea Purpurea." *Planta* 213 (2): 192–98. https://doi.org/10.1007/s004250000503.

Sevos, Jessica, Anne Grosselin, Jacques Pellet, Catherine Massoubre, and Denis Brouillet. 2013. "Grasping the World: Object-Affordance Effect in Schizophrenia." *Schizophrenia Research and Treatment* 2013: 531938. http://dx.doi.org/10.1155/2013/531938.

Shapiro, Lawrence. 2019. *Embodied Cognition.* 2nd ed. London and New York: Routledge.

Shaw, Robert E., M. T. Turvey, and William M. Mace. 1982. "Ecological Psychology. The Consequence of a Commitment to Realism." In *Cognition and the Symbolic Processes*, edited by Walter B. Weimer, and David S. Palermo, vol. 2, 159–226. Hillsdale, NJ: Lawrence Erlbaum Associates, Inc.

Stoffregen, Thomas A. 2003. "Affordances as Properties of the Animal-Environment System." *Ecological Psychology* 15 (2): 115–34. https://doi.org/10.1207/S15326969ECO1502_2.

Stolarz Maria. 2009. "Circumnutation as a Visible Plant Action and Reaction: Physiological, Cellular and Molecular Basis for Circumnutations." *Plant Signal Behaviour* 4 (5): 380–7. https://doi.org/10.4161/psb.4.5.8293.

Trewavas, Anthony. 2005. "Green Plants as Intelligent Organisms." *Trends in Plant Science* 10 (9): 413–19. https://doi.org/10.1016/j.tplants.2005.07.005.

Trewavas, Anthony. 2014. *Plant Behaviour and Intelligence.* New York, NY: Oxford University Press.

Turvey, M. T. 1992. "Affordances and Prospective Control: An Outline of the Ontology." *Ecological Psychology* 4 (3): 173–87. https://doi.org/10.1207/s15326969eco0403_3.

Turvey, M. T., Robert E. Shaw, E. S. Reed, and William M. Mace. 1981. "Ecological Laws of Perceiving and Acting: In Reply to Fodor and Pylyshyn." *Cognition* 9 (3): 237–304. http://dx.doi.org/10.1016/0010-0277(81)90002-0.

Wilson, Margaret. 2008. "How Did We Get from There to Here? An Evolutionary Perspective on Embodied Cognition." In *Handbook of Cognitive Science*, edited by Paco Calvo, and Antoni Gomila, 373–93. San Diego, Oxford and Amsterdam: Elsevier. https://doi.org/10.1016/B978-0-08-046616-3.00019-0

Yokawa, Ken, and František Baluška. 2018. "Sense of Space: Tactile Sense for Exploratory Behavior of Roots." *Communicative & Integrative Biology* 11 (2): 1–5. https://doi.org/10.1080/19420889.2018.1440881.

Zipoli Caiani, Silvano, and Gabriele Ferretti. 2017. "Semantic and Pragmatic Integration in Vision for Action." *Consciousness and Cognition* 48: 40–54. http://dx.doi.org/10.1016/j.concog.2016.10.009.

Part 4

Consciousness

9 Making Sense of Plant Sense

Brian Key and Deborah J. Brown

9.1 Introduction

The 1986 musical film *Little Shop of Horrors* features a carnivorous and sentient human-flesh-eating plant called Audrey II that brings fame and misfortune to a struggling florist shop. Audrey II can speak and demands that Seymour, the shop owner, "feed me all night long." The dumbfounded Seymour recoils—"You're a plant, an inanimate object." Audrey II then hollers "if I can talk, if I can move, who says I can't do anything I want." Later, we learn that Audrey II is an alien, a so-called "mean green mother from outer space." "That explains all the craziness", thinks the audience, "It's not that a plant is thinking or conscious; it's just that we've been invaded by aliens!" But either way, while we entertain the fiction, there is a suspension of the usual form-to-function constraint that is the hallmark of biological explanation. A sufficiently intelligent being could, we surmise, wreak havoc by assuming the form of a carnivorous pot plant.

Shortly before this cult film hit the big screen, the philosopher, David Lewis, concocted a similar thought experiment in which a Martian alien lacking a human nervous system is sentient because its water-filled cavities—much like the vascular bundles in some plants—enable it to feel sensations such as pain (Lewis 1980). Lewis' Martian is intended to illustrate the truth of multiple realisability—the idea that subjective experiences like pain could occur (and probably do occur across different species or individuals within the same species) in different physical forms. In another nearby possible world, Lewis deposits a madman who is in a neural state equivalent to the neural state we are in when in excruciating pain but because that state is wired up differently, it is caused by mild tickles and other innocuous stimuli and produces effects that normally would not be associated with being in pain. The madman's pain is not correlated with aversive behaviour and indeed stimulates him to concentrate on completing difficult mathematical proofs. Where Martian pain poses problems for any neat reduction of pain to a specific type of neural state typical of

DOI: 10.4324/9781003393375-14

humans, the madman is intended to show that the functional role of pain typical of humans is equally negotiable.

The functional role of pain in humans is, however, indirectly relevant to understanding what mental state the madman is in. The madman is in pain, but it is atypical pain, and this is okay because he is a member of a population (humans) in which that physical state normally has the function associated with pain. There is something to be said for this move. We regard people with diseased kidneys as still having kidneys not because their kidneys function to extract waste from their blood but because that structure in the human population usually performs that function. Nonetheless, there are all sorts of problems with Lewis' solution. Were the sub-population of madpeople to reach 50% during a madpain episode, would the madman suddenly cease to be in pain despite no change in the phenomenology of the experience? If function rather than phenomenology determines what a sentient state is, and madpain functions like an algorithm for transforming tickles into mathematical proofs, why isn't that its function rather than pain in accordance with functionalist criteria? Problems similar to those which arise for madpain could thus also arise for what we might call here mad kidney disease. Suppose that the structure which in humans is a kidney and typically extracts waste from the blood in mad people does something entirely different. Perhaps it detects skin-damaging UV levels. Would we say this structure is a dysfunctional kidney or a functional UV detector? Should it matter what percentage of the population the mad kidney people make up? If the identity of the structure as a kidney could change because of an extrinsic demographic shift without any internal alteration, we might be inclined to think, following the functionalist's logic, that the structure was no longer a kidney at all.

Structure matters. Function matters. And for the purposes of biological explanation, the two must be connected, the structure explaining how the function is able to be performed. The dissociations between the two we make in fiction and in thought experiments distract us from the fact that in nature it is their interdependence that counts. This brings us to the matter of plant sentience.

In what follows, we question not so much the fact that advocates of plant sentience rely on arguments from analogy, which are more problematic the more they ignore important dissimilarities between cases. To make informed judgements about a non-human organism's capacity for sentience, arguments from analogy are unavoidable because attributions of sentience to species incapable of self-reports are always going to depend on analogising from what we discover first about human sentience. Rather, we question a chief assumption upon which such arguments depend, namely, that the function of subjective experiences is to cause adaptive behaviour. We label this the "causal assumption." If it turns out that the causal role of subjective experiences is not to cause behaviour, as we propose to argue here, then inferences to the best explanation about the potential of plants to feel based on the causal assumption are dead in the watering can. This, in turn, has consequences for our confidence in the multiple realisability assumption and the cogency of thought experiments dissociating structure from function.

9.2 Neurocentrism v. Anthropocentrism

It helps us on our journey to accepting the dissociation between structure and function in Audrey II and Lewis' Martian that they talk to us about their feelings. We rightly trust self-reports of pain from other humans and linguistic ability since Descartes has been thought to be the principal marker of thought and consciousness. But with increasingly sophisticated language models such as ChatGPT-4, the sufficiency of this criterion is in doubt. So far generative AI is giving us a straight-up emphatic "no" in answer to the question whether it is sentient. When asked, ChatGPT-4 replied:

No, I am not sentient. I am a computer program running on a powerful machine, and I process and generate text based on patterns in the data I was trained on. I do not possess consciousness, self-awareness, emotions, or subjective experiences like a sentient being. My responses are generated based on the input I receive and the patterns I've learned from the text data I was trained on.

Turns out that ChatGPT-4 has a neurocentric bias. When asked what capacity it lacks to feel pain, it responds:

The capacity I lack to feel pain is a complex combination of biological, neurological, and emotional attributes that are present in living organisms. Pain is a subjective experience that involves the perception of harm or danger and the emotional response to that perception. It requires a nervous system, specialized receptors (such as nociceptors), and a brain to process and interpret the signals from those receptors. As a computer program, I lack a physical body, a nervous system, and the sensory and emotional processes necessary to experience pain. My functioning is based on algorithms and data processing, and I do not possess the necessary components for sensory perception or emotional experiences like pain.

It is important to distinguish neurocentrism—the idea that feelings like pain only emerge within sufficiently complex neural architectures—from anthropocentrism—the idea that feelings like pain are essentially human characteristics. Calvo, Baluška and Trewavas (2021) tend to blur this distinction in their defence of plant consciousness. They argue that evidence for plant consciousness should not be sought from studies of the nature of human consciousness. Rather, we should begin by defining the "primary biological function of consciousness" accepting that how it is realised morphologically will differ between organisms. (This is multiple realisability.) The function of consciousness is recognised by the way a conscious experience helps an organism deal with the outside world—e.g., avoiding a predator, seeking nourishment, etc. (This is the causal assumption.) Even they, however, inadvertently slide back to human consciousness to capture the function of consciousness: "because consciousness [here, they are referring to human consciousness] clearly influences behaviour, it has a biological function. And because it can influence behaviour, it must function as input data to a process or processes that determine behaviour." But their functional anthropocentrism—starting with a notion of consciousness defined by its biological function in humans—is supposedly acceptable, whereas neuro-anthropocentrism—starting with the structures that in humans explain the function of consciousness—is not. Calvo, Baluška and Trewavas (2021) are clear that the function of consciousness is to cause adaptive behaviours and contend that consciousness is thus itself adaptive and subject to selection pressures during evolution. Consciousness improves an organism's chances of survival in the wild. More recently, Segundo-Ortin and Calvo (2023) describe how the disruptive effects of anaesthesia on electrophysiological processes and plant behaviour are potential indicators of plant sentience. They argue against the idea that plant behaviour is reactive and hardwired and instead stress the flexible nature of plant cognition as a hallmark of sentience. An overriding theme is that there is no constraint on the kinds of structures that can realise consciousness once the function of sentience is fixed.

Freed from any obligation to restrict attributions of sentience to organisms with a brain, advocates of plant sentience feel licenced in relying upon "neuronlike" vascular structures in their attributions of sentience to plants (Marder 2012, 2014; Gagliano 2015; Pelizzon and Gagliano 2015; Calvo 2017; Calvo, Sahi, and Trewavas 2017; Gagliano 2017; Trewavas 2017; Baluška and Mancuso 2018; Gagliano 2018; Yokawa and Baluška 2018; Baluška and Reber 2019; Parise, Gagliano, and Souza 2020). In arguing for the legitimacy of designating the field "plant neurobiology", Segundo-Ortin and Calvo (2023, 15–16) argue that "restricting the term 'neurobiology' too tightly to neurons risks missing many molecular-level functional similarities between animal and plant signaling systems and substrates", in particular, the presence in plants of action potentials (APs) and chemicals also found in neurotransmitters (e.g., acetylcholine, glutamate, dopamine, histamine, noradrenalin, serotonin, and g-aminobutyric acid (GABA)). The fact that APs and the presence of such chemicals are also correlated with non-conscious brain activity in humans (Sahraie et al. 1997) is not considered. Regarding behavioural criteria, there is widespread agreement among advocates that plants exhibit "goal-directed behaviors", "make decisions", "remember" and "communicate", via either airborne volatile organic compounds (VOCs) (Baldwin 2010; Baldwin et al. 2006) or ground vibrations (Gagliano et al. 2012), which makes their sensory systems more sophisticated than devices like thermostats that detect changes in their environments. Plants "self-recognise", "recognise kin", "navigate", "integrate information", "anticipate", "learn", "respond to anaesthetics" and so on, leading Gagliano (2017) to declare the existence of nothing short of a "vegetal mind." The vascular systems of plants are heralded as the "plant-specific neuronal systems needed for the execution of adaptive goal-directed behaviors" whereas the root apices are "rich in sensory systems", which together create "neuronal networks" (Baluška and Mancuso 2013, 2020) that are sufficient to support sentience.

We have previously argued (Brown and Key 2021b) that these are weak analogies that rely on anthropomorphising uses of intentional terminology where mechanosensory descriptions would be more justified, and that, if correct, such analogies would justify the attribution of sentience to long-distance telephone cables and microcircuits. We further argued that arguments for plant sentience tend to be rooted in questionable gradualist assumptions—in particular, that evolutionary continuity entails anti-emergentism (i.e., that properties like consciousness could not emerge suddenly in evolution) and that shared ancestry of plants and animals is evidence of linked evolution of common morphological features rather than convergent evolution. Drawing on both naturally emergent features (e.g., the emergence of the facial muscle responsible for puppy dog eyes) and genetic experimentation (e.g., the generation of a novel axon tract in zebrafish), we showed how small changes in genotype can generate new emergent properties. The gradual evolution of genotypes is consistent, in other words, with the non-gradual evolution of phenotypes. There is no reason thus to suppose, as William James (1890, 152) did, that for "evolution to work smoothly, consciousness in some shape must have been present at the very origin of things."

We also took issue there with the too-quick dismissal of semantic objections to the idea of plant-specific pain and consciousness more generally. A common retort to arguments against attributing to non-human species X the capacity for pain or some other subjective experience is to say that X may have its own type of pain or subjective experience—e.g., "X pain", for any X. This rests on a mistake. What, we ask, would "X pain" have to do with pain if it is not pain? Defenders of plant sentience are apt to dismiss semantic objections as "sterile arguments" (Calvo and Trewavas 2020) derailing progress in the science of consciousness (Chamovitz 2018;

Robinson, Draguhn, and Taiz 2020; Maher 2020). But the semantic objection is not to be trivialised. The metaquestion is whether the question "Is 'X pain' pain?" is like the question "Is a hotdog a sandwich?" or more like the question "Is a donkey's tail a leg?" We can go either way on both these questions, but not without potentially causing much confusion. Less so in the case of the hotdog—the dog bit is between chunks of bread, so it is more like a sandwich than a Caesar salad is. But while we can call a donkey's tail a leg, we don't thereby make it one and only cause confusion if we do. If the hotdog is a sandwich, it really is a sandwich and we can explain why, but we cannot similarly explain why the donkey's tail would be a leg. We might be able to explain why we *call* it so—perhaps the tail looked from a distance like a leg or we're using "leg" ambiguously as Humpty Dumpty does when he uses "glory" to mean "a nice knockdown argument." But this does not explain why that would make the tail a leg. The dilemma in the case of the question "Is X pain, pain?" is that if we answer "yes", we don't need the nomenclature of calling it "X pain" at all—it is just pain in X, the same as pain in humans or rats—whereas if we answer "no", we have no basis for thinking that "X pain" is anything conscious at all. If we have no way of saying what it is like to feel X pain (because it is not pain we are talking about after all), we have no way of knowing that there is anything it is like to feel X pain at all. Resorting to behavioural or electrophysiological evidence will not help since such evidence is insufficient to disconfirm the hypothesis that the organism is undergoing non-conscious nociception with no subjective feeling at all.

Much has since been written in opposition to the proposal that plants are sentient and to the appropriateness of "plant neurobiology" as a designation for a field of study of organisms lacking neurons (see responses to Segundo-Ortin and Calvo's (2023) target article). We will not rehearse those objections here. Rather, we take issue with the hitherto overlooked assumption that the function of feelings is to cause adaptive behaviours, which advocates of plant sentience take to licence their inference from the electrophysiological activity in plants causing behaviour to its constituting a felt experience. This, we argue, is a false assumption and thus the evidence for establishing sentience goes beyond the presence of electrophysiologically modified adaptive behaviours. If, instead, felt experience depends on cortical-level processing of information, absence of these architectures in plants will constitute evidence of an absence of sentience (Brown and Key 2021a).

9.3 What Exactly Is the Function of Sensory Experiences?

Calvo, Baluška and Trewavas (2021) represent a widespread consensus that the function of subjective experience is to control behaviour, pain being a case in point. Many scientific studies or reviews of pain proclaim that the function of pain is to protect the organism from harm, an assumption with which lay people would readily agree (Nash 2005). What follows are some typical statements: "pain constitutes an alarm that ultimately has the role of helping to protect the organism" (Le Bars, Gozariu, and Cadden 2001); "the sensation of pain protects humans (and other species) from the tissue-damaging effects of dangerous stimuli, and appears to be critical for survival of the organism" (Nagasako, Oaklander, and Dworkin 2003); "pain is a fundamental experience that promotes survival" (Atlas 2021); "pain is a highly salient signal, and acute pain in many ways protects us from injury and harm" (Sun et al. 2023). The basic idea is that pain has an evolutionary function for an organism and is not an epiphenomenon (Kolodny, Moyal, and Edelman 2021). That being so, the idea that pain is protective seems to come principally from first-person experience—e.g., one learns not to touch a hot stove so as not to experience pain. Mancini, Zhang and Seymour (2022) recognise this when they state that "the main function of the pain system is to minimise harm and, to achieve this goal, it needs to learn to predict forthcoming pain."

Pain is a primary example of the kind of sentience of which plants, according to proponents of plant consciousness, are supposedly capable. The idea was initially canvassed by Baluška and Mancuso (2009). More recently, Baluška and Yokawa (2021) note that "with respect to pain perception, it is obvious that plants do not perceive noxious stimuli in the same way as animals and humans." Nonetheless, they believe the function of pain for plants is to enable them "to survive and adapt effectively, (since) plants still need to be able to recognise what is dangerous." This survival function is supported by Calvo's claim that "plants can not only learn and memorize, but also make decisions and solve complex problems" (Calvo 2016).

This idea that pain is a teaching signal used in learning to actively avoid pain (Koppel et al., 2023) is consistent with what is known about physiological responses to noxious stimuli in humans. The subjective experience of pain is too slow to act as a first line of defence against noxious stimuli—the injury and harm is already done to the body before pain is felt. The first line of defence against noxious stimuli to the limbs is the withdrawal or flexion reflex (Sherrington 1910). Electric shocks to the upper limb digits elicit reflex limb motor activity in about 100 milliseconds, which is considerably faster than voluntary responses (which depends on the conscious perception of pain from the electric shock). The voluntary, conscious response occurs at a latency of about 200 milliseconds (Floeter et al. 1998; Peterson et al. 2014). Thus, pain is not a very good first line of defence for preventing bodily harm—relying on pain alone would result in serious tissue damage. It is commonly believed that pain is essential for protection and survival because patients with congenital insensitivity to pain typically suffer from horrific injuries of the limbs and have a reduced lifespan. However, what is typically misunderstood is that these patients lack pain because of the absence of nociception either because of a lack of sensory fibres or a channelopathy (Nagasako, Oaklander, and Dworkin 2003; Cox et al. 2006; Weisman, Quintner, and Masharawi 2019). Their peripheral nerves do not transmit noxious signals to the spinal cord and then to the brain and it for this reason they lack pain. The main reason for the shortened lifespan turns out not to be the absence of nociception and pain per se, but rather an increased prevalence to partake in risky behaviour. These patients do not learn to avoid risky behaviours that can lead to serious bodily harm. Could the function of pain then be to teach us how to manage risk? It is not possible to determine in these patients whether the underlying teaching signal is either nociception or pain since they lack both. Intuitively, subjects that experience pain believe pain must be the teaching signal since it is pain that one seems to seek to avoid. However, to answer the question which of pain or nociception is the teaching signal, we really need to assess patients with normal nociception but who congenitally lack pain. Unfortunately, such individuals do not exist.

It is well established that associative learning does not require awareness of the conditioned stimulus. Wong et al. (1997) showed that images of angry faces, when presented below the level of conscious awareness, could be conditioned to elicit conditioned physiological responses by training with electric shock in humans. This result is consistent with the important function of pain as necessary for associative learning but what about when the unconditioned stimulus is below the level of conscious awareness? It seems incredible that learning could occur when the aversive signal is nonconscious—why else would one avoid a stimulus if it wasn't experienced as something awful? If such learning were possible, it would negate the evolutionary function of pain as a survival factor and seem to make pain a mere epiphenomenon. Moreover, it would spell the end of plant pain. How do you show that the experience of electric shock as pain is not necessary for associative learning? It is not enough to use electric shock as an unconditioned stimulus in an animal one considers as having a simple nervous system, such as Aplysia or jellyfish. To do so only means you must first assume that these animals can or cannot feel pain.

9.4 What Has Been Learned from a Slug

Given that the neural circuitry underlying aversive associative learning in Aplysia is known (Hawkins 2019), we can assess the likelihood that this circuitry has the capacity to support the experience of pain or not—an approach we previously applied to worms (Zalucki, Brown, and Key 2023). Aversive associative learning in Aplysia typically involves conditioning the siphon withdrawal reflex, which is induced by weak touch

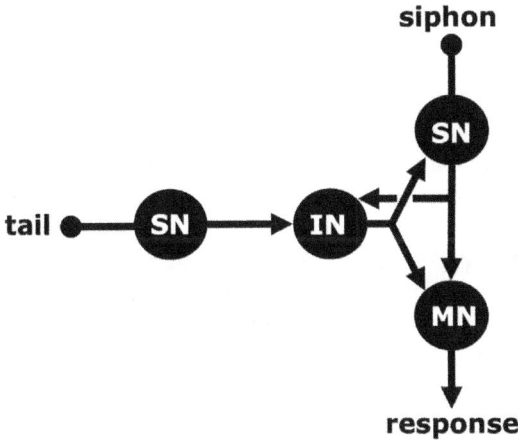

Figure 9.1 Circuit diagram of aversive associate learning involving the siphon reflex in Aplysia. SN, sensory neuron; IN, interneuron; MN, motor neuron.

(i.e., a conditioning stimulus such as a water jet) to the siphon, to the unconditioned stimulus (electric shock to the tail). This reflex is driven by a monosynaptic synapse between touch sensory neurons and motor neurons in conjugation with serotonin expressing interneurons (Fig. 9.1; Cleary, Byrne, and Frost 1995; Hawkins 2019). Electric shock to the tail activates nociceptive sensory neurons which synapse on interneurons that modulate the excitability of both touch sensory neurons and motor neurons. The siphon sensory neurons also converge on the same interneurons innervated by the tail sensory neurons. After conditioning, the withdrawal reflex elicited by weak touch alone is much stronger than prior to conditioning. This conditioning is dependent on the facilitatory effects of the interneurons on the siphon reflex (Antonov et al. 2003).

If this aversive associative learning was driven by the subjective experience of pain, then this pain must arise within the simple feedforward circuit between the tail sensory neuron and the facilitatory interneurons. Since there is convergence and integration of sensory information from distinct sources (touch and nociception) onto a single interneuron, this interneuron cannot be the site where a discrete sensation such as pain or gentle touch is generated. Given that neural activity in isolated sensory neurons has no possibility of generating a subjective experience (otherwise, every cultured sensory neuron would be conscious), it becomes clear that aversive associative learning in Aplysia is instead driven by nonconscious neural activity. Thus, the idea that an organism's survival from harm arising from noxious stimuli is dependent on the subjective experience of pain is a misleading claim.

9.5 What Has Been Learned from a Fruit Fly

It might be objected that Aplysia is a special case and not like other organisms. Perhaps other invertebrates with more complex nervous systems—such as Drosophila—do require pain for aversive associative learning. A common form of aversive associative learning examined in Drosophila involves the learned association between a neutral olfactory stimulus (conditioned stimulus) and an electric shock (unconditioned stimulus) which causes the conditioned response (avoidance behaviour) when the olfactory stimulus alone is presented. This associative learning occurs in the mushroom body of the Drosophila brain and is dependent on dopamine-expressing interneurons (referred to as DANs) that—like the serotonin interneurons in Aplysia—integrate information from different sources to facilitate conditioned motor output (Li et al. 2020; Fig. 9.2). The converging neural activity from olfactory and noxious sensory stimuli onto DANs creates a unique internal brain state of Drosophila (Siju, De Backer, and Grunwald Kadow 2021). While these neurons are far more numerous than the serotonin-expressing facilitatory interneurons in Aplysia, they are situated in a similarly designed canonical circuit. DANs increase their activity in response to the converging sensory inputs and train the circuit to generate a repulsive response to a previously neutral odour.

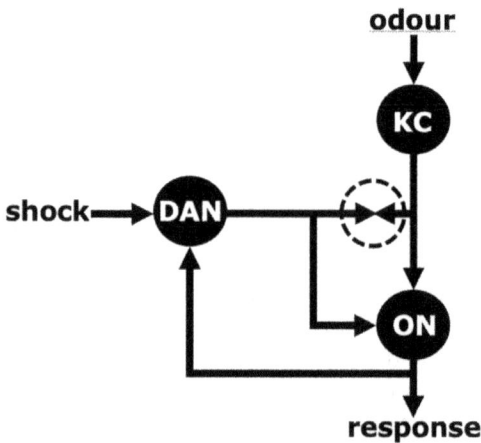

Figure 9.2 Canonical circuit for aversive associative learning in the Drosophila mushroom body. Based on Li et al. (2020). Note that DAN inhibits the KC-MBON pathway to reduce what is normally an attractive response (Owald et al. 2015). Also, the reciprocal synapse (dashed circle) between KC and DAN allows feedforward to DAN (Cervantes-Sandoval et al. 2017). DAN, dopaminergic neuron; ON/MBON, mushroom body output neuron; KC, Kenyon cells.

There are many observations that illustrate DANs do not and cannot generate the subjective experience of pain. (1) Prior to associative learning, DANs respond, albeit weakly, to olfactory stimuli as well as to electric shock when each is presented separately (Riemensperger et al. 2005). Given that neutral olfactory stimuli do not generate aversive responses when presented alone, it is unlikely that these neurons generate pain. (2) DANs show increased neural activity with delivery of electric shock. However, after training, these neurons show similar increased activity following presentation of the conditioned olfactory stimulus alone (Riemensperger et al. 2005). We know from human experience that there is no feeling of pain after aversive associative learning when only the conditioned stimulus is presented. Thus, neural activity in DANs is not pain, but it is instead a nonconscious teaching signal. (3) During aversive associative learning, DANs are integrating both olfactory and noxious inputs and are therefore unlikely to represent the discrete subjective experience of either pain or smell. (4) DANs receive converging information from multiple aversive sensory modalities including electric shock, heat, and bitter taste (Galili et al. 2014; Li et al. 2020). Thus, DANs cannot separately represent the subjective experience of each of these stimuli. (5) Drosophila exhibit second-order aversive conditioning which has been previously promoted as evidence of subjective experience (Birch, Ginsburg, and Jablonka 2020). First-order conditioning occurs when a neutral olfactory stimulus is conditioned by electric shock to produce an aversive response. Second-order conditioning occurs when the first olfactory stimulus is then paired with a second neutral olfactory stimulus during subsequent training so that the aversive response is now transferred to that second-order stimulus. This second-order conditioning is driven by a simple circuit linking the conditioned response of mushroom body outputs to a second pool of DANs by a specific intermediatory interneuron (Jürgensen et al. 2023; Yamada et al. 2023). These DANs then condition a population of Kenyon cell-mushroom body output neurons activated by the second olfactory stimulus so that they now also drive an aversive response. Given that these second-order DANs are not activated by electric shock, pain is clearly not driving second-order aversive associative learning. It is therefore a mistake to think that second-order conditioning necessarily indicates the presence of subjective experience. (6) Electric shock-induced activity in DANs can be replaced by mere optogenetic activation of these cells with laser light (notably this activity lacks any of the normal temporal patterning that would occur during natural processing) (Hige et al. 2015). Given that this artificial activation of DANs is sufficient to drive aversive associative learning, there is no need to think that it is necessary for learning that electric shock causes pain. (7) Activation of a single DAN is sufficient to drive aversive associative learning (Hige et al. 2015), which dismisses any possibility that local networks of DANs are somehow generating subjective experience. In this respect, the circuitry in Drosophila is very similar to that in Aplysia. (8) The computations executed by DANs are embedded in the local circuitry of the mushroom body and only function locally as teaching signals. This restricted connectivity negates their ability to generate system awareness that has been considered necessary for subjective experience (Key and Brown 2018; Key et al. 2021; 2022). In summary, evidence from aversive associative learning paradigms in Drosophila do not support the premise that the subjective experience of pain is necessary for protection from harm.

9.6 What Has Been Learned from Mammals

Aversive associative learning in rodents using olfactory cues (conditioned stimuli) and electric shocks (unconditioned stimuli) involves neural circuits in brain regions (Perisse, Miranda, and Trouche 2023) such as the amygdala, ventral tegmental area, thalamus, hippocampus and cerebral cortex, which have been implicated in contributing to subjective experience in mammals, including humans. To address the question of whether subjective experience of pain is needed to avoid harm in mammals, a model system is needed where the influence of subjective experience can be definitively removed. The spinal cord isolated from the brain provides such a model. We know from humans with complete spinal cord lesions that such preparations lack conscious experience. Paraplegic patients can exhibit normal withdrawal reflex responses to noxious stimuli applied to their paralysed limbs without consciously experiencing any feelings of pain (Defrin et al. 2007). Because of this, we can dismiss any suggestion that an isolated spinal cord is conscious. Moreover, direct electrical stimulation of the somatosensory cortex in the human brain isolated from the spinal cord (*i.e.*, in a tetraplegic patient) produces normal somatosensation and reveals that the cortex and not the spinal cord is the site of conscious experience (Flesher et al. 2016). Rats with thoracic spinal transections exhibit aversive associative learning in the isolated lumbosacral cord in the absence of conscious experience of either the electric shock or proprioceptive cues from the hindlimb (Grau 2014). The animal is positioned so that when one hindlimb extends past a critical limb position (proprioceptive cues from the limb are the conditioned stimulus), it activates an electric shock (unconditioned stimulus) to the paw of that limb. This shock elicits a flexion withdrawal reflex that lasts less than five seconds and temporarily extinguishes the shock. Within 20 shocks, the animal can continuously maintain the limb in a flexed position (conditioned response) for up to 50 seconds. These results reveal that the mammalian spinal cord can learn to reduce harm without the need for conscious control or subjective experience of pain.

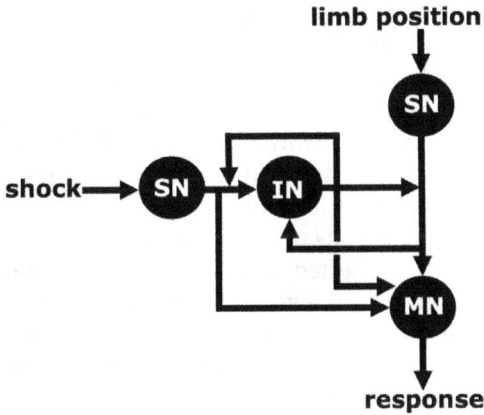

Figure 9.3 Circuit diagram for aversive associate learning in the isolated mammalian spinal cord. The connectivity in this circuit reflects that present in Apylsia (Fig. 9.1) and Drosophila (Fig. 9.2). SN, sensory neuron; IN, interneuron; MN, motor neuron.

Aversive associative learning has also been demonstrated in paralysed dogs using electric shock to the tail as the conditioned stimulus (the intensity of this shock fails to cause contraction of a hindlimb flexor muscle; Shurrager and Culler 1940). The unconditioned stimulus was an electric shock to the paw which caused a strong muscle contraction. After training, the conditioned stimulus applied alone was able to elicit flexor muscle contractions demonstrating aversive associative learning in the absence of conscious experience. Similar results have been reported for aversive associative learning in paralysed cats (Fitzgerald and Thompson 1967).

Preliminary studies have begun to interrogate the circuitry underlying spinal learning using the isolated mouse spinal cord (Lavaud et al. 2022). While much work remains to be done, a specific subpopulation of inhibitory interneurons expressing the Ptf1a gene in the dorsal spinal cord plays an important role. The circuitry teased out to date (Fig. 9.3) looks like the canonical circuits driving the same learning in Aplysia (Fig. 9.1) and Drosophila (Fig. 9.2). The Ptf1a interneuron integrates neural inputs from the proprioceptive and nociceptor sensory neurons which are then fed-forward to regulate the proprioceptive-motor neuron pathway that generates the conditioned response.

9.7 The Biological Function of Pain

As we have seen from the analysis of aversive associative learning, pain need not be a direct motivating agent. Instead, aversive behavioural responses can be controlled by interneurons integrating converging inputs from noxious stimuli and other sensory modalities. These interneurons then modify the synaptic efficiency of circuits linking the relevant sensory cues and the motor output pathways to generate the appropriate behaviour.

Cognition is classically considered to involve the manipulation of symbols (Casamajor 1929; Fodor 1975; van Gelder 1995). Casamajor (1929) called these symbols the "coinage of the mind." Although it is well known that there are both nonconscious and conscious types of cognition (Cleeremans 2006; 2014), it is often neglected that conscious cognition selectively involves the manipulation of conscious symbols, and that subjective experience may be one such symbol. Conscious cognitive processes can employ symbols for reasoning and planning. These symbols capture semantic information, i.e., information that is meaningful for a system. In the case of vision, an image may contain semantic features like faces that are highly meaningful for primates and can be considered to act as symbols. These features are preferentially processed in select regions of the inferior temporal lobe such as the fusiform facial area (Puce et al. 1995). We also consider subjective experience (e.g., pain) as being symbolic in the sense that it has semantic information and can act like visual features and linguistic symbols to assist in reasoning, planning, and executing actions. In this context, it makes sense that sensory stimulation should be consciously experienced since it is this quality that ensures its use by conscious cognitive processes. Moreover, it follows that sensory stimulation should feel like something rather than nothing so that a non-linguistic and cognitively conscious animal can reason, plan and execute actions using these experiences. We propose that subjective experience, such as pain, is a conscious *explanation* of an animal's behaviour to its cognitively conscious brain, i.e., "explanation" in a weak sense of making sense of an organism's behaviour to itself. Why did I withdraw my hand from the stove? Because it hurts. In this manner, it acts as a heuristic that automatically explains to a reasoning brain why it is behaving in the way it does. Since the explanation/pain is subsequent to the nocifensive behaviour, it is too late to perform this causal role and the explanation is thus false. This does not mean that the subjective experience is itself a fiction—it is real and enables us to predict and plan to avoid circumstances which threaten bodily integrity.

We propose that the function of subjective experience is to explain to the cognitively conscious brain why it behaves as it does. This hypothesis also explains why sensory stimulation evolved to feel like something rather than nothing since it is the inner form of communication of cognitively conscious organisms that lack symbolic language. A prediction of this hypothesis is that subjective experience would only exist in organisms that are cognitively conscious (which we refer to here as being knowingly aware of goal-directed neural processing) or that have the potential to become consciously cognitive (e.g., during development).

9.8 Plant Sense Does Not Make Sense

We now return to Calvo's quest for a primary function of subjective experience. We have shown here that subjective experience is not necessary for harm reduction or prevention; hence, just observing such behaviour does not necessarily imply that an organism is consciously experiencing feelings. Nor is the presence of electrophysiological activity sufficient to warrant attributions of sentience. We appreciate that plants have very unique sensory mechanisms (Arimura et al. 2000; Nagashima et al. 2019; Wu 2023; Yang et al. 2023) but not in any way that implies sentience. They detect changes in their environments and integrate information to adapt their behaviour to changing circumstances, but since neither of these functions implies consciousness, their sensing capabilities are different only in the degree of complexity, not kind from those of mechanical devices like the thermostat (Struik 2023). In our framework, subjective experience serves no function in the absence of conscious cognition. Given that there is no evidence that plants possess conscious cognition, there is no need for them to evolve what would be expected to be an energetically expensive property such as subjective experience. Ipso facto, plants do not feel or have feelings.

9.9 Conclusion

We began with thought experiments based on the dissociation of structure and function for a subjective experience like pain. Audrey II and Martian pain are examples of beings which appear to satisfy the causal assumption (that it is subjective experience causing their behaviour) without that involving anything remotely like the cortical structures typical of humans. Both have vascular structures similar to those of plants. We are helped in our acceptance of the conclusions of these thought experiments by the way in which they are described—we are *given* the fact that these are intelligent, talking, self-moving aliens. But if the causal assumption is false, we are not warranted in inferring either that sentience is multiply realisable or the conclusions of the usual thought experiments. We'd need more information than that an organism's nocifensive behaviour is mediated by some internal electrophysiological activity. Mad pain suspends the causal or functional criterion retaining a minimal version of the structural condition. In our terms, mad pain does not play its usual explanatory role; it does not constitute a form of conscious cognition, or, if it does, it answers a different question, namely, why the madman is producing mathematical proofs. But if it at all feels like pain, it is because it minimally satisfies the structural condition and would be cognitively accessible to the madman, not because of any extrinsic relation the madman bears to normal human sufferers. And if it is pain, it is the same as human pain or rat pain, not a distinct kind of pain, "madpain." Again, from a comparative neuroana-tomical perspective, the conditions for madpain are too underdescribed to warrant a conclusive answer to the question whether this counts as pain. The cases presented for plant sentience bear many similarities to these thought experiments, forgoing for the sake of convenience a commit-ment either to the precise structural conditions without which we cannot draw any conclusions about whether an organism is feeling anything, or a commitment to precision in defining the function of sentience. This is not advancing the science of sentience but miring it in confusion from which it would be difficult to return.

Acknowledgements

This work was supported by an Australian Research Council Discovery Grant DP200102909.

References

Antonov, Igor, Irina Antonova, Eric R. Kandel, and Robert D Hawkins. 2003. "Activity-Dependent Presynaptic Facilitation and Hebbian LTP are Both Required and Interact During Classical Conditioning in Aplysia." *Neuron* 37 (1): 135–47. https://doi.org/10.1016/s0896-6273(02)01129-7.

Arimura, Gen-ichiro, Rika Ozawa, Takeshi Shimoda, Takaaki Nishioka, Wilhelm Boland, and Junji Takabayashi. 2000. "Herbivory-induced Volatiles Elicit Defence Genes in Lima Bean Leaves." *Nature* 406 (6795): 512–15. https://doi.org/10.1038/35020072.

Atlas, Lauren Y. 2021. "A Social Affective Neuroscience Lens on Placebo Analgesia." *Trends in Cognitive Sciences* 25 (11): 992–1005. https://doi.org/10.1016/j.tics.2021.07.016.

Baldwin, Ian T. 2010. "Plant Volatiles." *Current Biology* 20: R392–7. https://doi.org/10.1016/j.cub.2010.02.052.

Baldwin, Ian T., Rayko Halitschke, Anja Paschold, Caroline C. von Dahl, and Catherine A. Preston. 2006. "Volatile Signaling in Plant-plant Interactions: 'Talking Trees' in the Genomics Era." *Science* 311: 812–15. https://doi.org/10.1126/science.1118446.

Baluška, František, and Stefano Mancuso. 2009. "Plant Neurobiology: From Sensory Biology, Via Plant Communication, to Social Plant Behavior." *Cognitive Processing* 10: 3–7. https://doi.org/10.1007/s10339-008-0239-6.

Baluška, František, and Stefano Mancuso. 2013. "Microorganism and Filamentous Fungi Drive Evolution of Plant Synapses." *Frontiers in Cellular and Infection Microbiology* 3: 44. https://doi.org/10.3389/fcimb.2013.00044.

Baluška, František, and Stefano Mancuso. 2018. "Plant Cognition and Behavior: From Environmental Awareness to Synaptic Circuits Navigating Root Apices." In *Memory and Learning in Plants*, edited by František Baluška, Monica Gagliano, and Guenther Witzany, 51–77. Cham: Springer. https://doi.org/10.1007/978-3-319-75596-0_4.

Baluška, František, and Stefano Mancuso. 2020. "Plants, Climate and Humans: Plant Intelligence Changes Everything." *EMBO Reports* 21 (3): e50109. https://doi.org/10.15252/embr.202050109.

Baluška, František, and Arthur Reber. 2019. "Sentience and Consciousness in Single Cells: How the First Minds Emerged in Unicellular Species." *BioEssays* 41 (3): 1800229. https://doi.org/10.1002/bies.201800229.

Baluška, František, and Ken Yokawa. 2021. "Anaesthetics and Plants: From Sensory Systems to Cognition-based Adaptive Behaviour." *Protoplasma* 258: 449–54. https://doi.org/10.1007/s00709-020-01594-x.

Birch, Jonathan, Simona Ginsburg, and Eva Jablonka. 2020. "Unlimited Associative Learning and the Origins of Consciousness: A Primer and Some Predictions." *Biology & Philosophy* 35: 1–23. https://doi.org/10.1007/s10539-020-09772-0.

Brown, Deborah J., and Brian Key. 2021a. "Is an Absence of Evidence of Pain Ever Evidence of Absence?" *Synthese* 199: 3881–902. https://doi.org/10.1007/s11229-020-02961-0.

Brown, Deborah J., and Brian Key. 2021b. "Plant Sentience, Semantics, and the Emergentist Dilemma." *Journal of Consciousness Studies* 28 (1–2): 155–83.

Calvo, Paco. 2016. "The Philosophy of Plant Neurobiology: A Manifesto." *Synthese* 193: 1323–43. https://doi.org/10.1007/s11229-016-1040-1.

Calvo, Paco. 2017. "What is it Like to be a Plant?" *Journal of Consciousness Studies* 24 (9–10): 205–27.

Calvo, Paco, František Baluška, and Anthony Trewavas. 2021. "Integrated Information as a Possible Basis for Plant Consciousness." *Biochemical and Biophysical Research Communications* 564: 158–65. https://doi.org/10.1016/j.bbrc.2020.10.022.

Calvo, Paco, Vaidurya Pratap Sahi, and Anthony Trewavas. 2017. "Are plants sentient?" *Plant, Cell & Environment* 40 (11): 2858–69. https://doi.org/10.1111/pce.13065.

Calvo, Paco, and Anthony Trewavas. 2020. "Physiology and the (neuro) Biology of Plant Behavior: A Farewell to Arms." *Trends in Plant Science* 25 (3): 214–16. https://doi.org/10.1016/j.tplants.2019.12.016.

Casamajor, Louis. 1929. "The Evolution of Symbols in the Development of Consciousness." *Archives of Neurology & Psychiatry* 22 (5): 865–72.

Cervantes-Sandoval, Isaac, Anna Phan, Molee Chakraborty, and Ronald L. Davis. 2017. "Reciprocal Synapses between Mushroom Body and Dopamine Neurons form a Positive Feedback Loop Required for Learning." *Elife* 6: e23789. https://doi.org/10.7554/eLife.23789.

Chamovitz, Daniel. A. 2018. "Plants Are Intelligent; Now What?" *Nature Plants* 4: 622–3. https://doi.org/10.1038/s41477-018-0237-3.

Cleary, Leonard J., John H. Byrne, and William N. Frost. 1995. "Role of Interneurons in Defensive Withdrawal Reflexes in Aplysia." *Learning & Memory* 2 (3–4): 133–51. https://doi.org/10.1101/lm.2.3-4.133.

Cleeremans, A. 2006. "Conscious and Unconscious Cognition: A Graded, Dynamic Perspective." In *Progress in Psychological Science Around the World. Vol I. Neural, Cognitive and Developmental Issues*, edited by Qicheng Jing, Mark R. Rosenzweig, Gery d'Ydewalle, Hucan Zhnag, Hsuan-Chih Chen, and Kan Zhange, 401–18. Hove: Psychology Press.

Cleeremans, Axel. 2014. "Connecting Conscious and Unconscious Processing." *Cognitive Science* 38 (6): 1286–315. https://doi.org/10.1111/cogs.12149.

Cox, James J., Frank Reimann, Adeline K. Nicholas, Gemma Thornton, Emma Roberts, Kelly Springell, Gulshan Karbani, Hussain Jafri, et al. 2006. "An SCN9A Channelopathy Causes Congenital Inability to Experience Pain." *Nature* 444 (7121): 894–8. https://doi.org/10.1038/nature05413.

Defrin, Ruth, Smadar Peleg, Harold Weingarden, Rafi Heruti, and Gideon Urca. 2007. "Differential Effect of Supraspinal Modulation on the Nociceptive

Withdrawal Reflex and Pain Sensation." *Clinical Neurophysiology* 118: 427–37. https://doi.org/10.1016/j.clinph.2006.10.015.

Fitzgerald, Linda A., and Richard F. Thompson. 1967. "Classical Conditioning of the Hindlimb Flexion Reflex in the Acute Spinal Cat." *Psychonomic Science* 8 (5): 213–14. https://doi.org/10.3758/BF03331626.

Flesher, S. N., Jennifer L. Collinger, Stephen T. Foldes, Jeffrey M. Weiss, John E. Downey, Elizabeth C. Tyler-Kabara, Sliman J. Bensmaia et al. 2016. "Intracortical Microstimulation of Human Somatosensory Cortex." *Science Translational Medicine* 8 (361): 361ra141. https://doi.org/10.1126/scitranslmed.aaf8083.

Floeter, Mary K., Christian Gerloff, Joshua Kouri, and Mark Hallett. 1998. "Cutaneous Withdrawal Reflexes of the Upper Extremity." *Muscle & Nerve: Official Journal of the American Association of Electrodiagnostic Medicine* 21 (5): 591–8. https://doi.org/10.1002/(SICI)1097-4598(199805)21:5<591::AID-MUS5>3.0.CO;2-3.

Fodor, Jerry A. 1975. *The Language of Thought*. vol. 5. Cambridge, MA: Harvard University Press.

Gagliano, Monica. 2015. "In a Green Frame of Mind: Perspectives on the Behavioural Ecology and Cognitive Nature of Plants." *AoB Plants* 7: plu075. https://doi.org/10.1093/aobpla/plu075.

Gagliano, Monica. 2017. "The Mind of Plants: Thinking the Unthinkable." *Communicative & Integrative Biology* 10 (2): e1288333.. https://doi.org/10.1080/19420889.2017.1288333.

Gagliano, Monica. 2018. "Inside the Vegetal Mind: On the Cognitive Abilities of Plants." In *Memory and Learning in Plants. Signaling and Communication in Plants*, edited by František Baluška, Monica Gagliano, and Guenther Witzany, 215–20. Cham: Springer. http://dx.doi.org/10.1007/978-3-319-75596-0_11.

Gagliano, Monica., Stefano Mancuso, and Daniel Robert. 2012. "Towards Understanding Plant Bioacoustics." *Trends in Plant Science* 17: 323–5. https://doi.org/10.1016/j.tplants.2012.03.002.

Galili, D. S., Kristina V. Dylla, Alja Lüdke, Anja B. Friedrich, Nobuhiro Yamagata, Jin Yan Hilary Wong, Chien Hsien Ho, et al. 2014. "Converging Circuits Mediate Temperature and Shock Aversive Olfactory Conditioning in Drosophila." *Current biology* 24 (15): 1712–22. https://doi.org/10.1016/j.cub.2014.06.062.

Grau, James W. 2014. "Learning from the Spinal Cord: How the Study of Spinal Cord Plasticity Informs Our View of Learning." *Neurobiology of Learning and Memory* 108: 155–71. https://doi.org/10.1016/j.nlm.2013.08.003.

Hawkins, Robert D. 2019. "The Contributions and Mechanisms of Changes in Excitability During Simple Forms of Learning in Aplysia." *Neurobiology of Learning and Memory* 164: 107049. https://doi.org/10.1016/j.nlm.2019.107049.

Hige, Toshihide, Yoshinori Aso, Mehrab N. Modi, Gerald M. Rubin, and Glenn C. Turner. 2015. "Heterosynaptic Plasticity Underlies Aversive Olfactory Learning in Drosophila." *Neuron* 88 (5): 985–98. https://doi.org/10.1016/j.neuron.2015.11.003.

James, William. 1890/1950. *The Principles of Psychology*. New York, NY: Dover.

Jürgensen, Anna-Maria, Felix-Johannes Schmitt, and Martin Paul Nawrot. 2023. "Minimal Circuit Motifs for Second-order Conditioning in the Insect Mushroom Body." *Frontier in Psychology* 14. https://doi.org/10.3389/fphys.2023.1326307.

Key, Brian, and Deborah J. Brown. 2018. "Designing Brains for Pain: Human to Mollusc." *Frontiers in Physiology* 9. https://doi.org/10.3389/fphys.2018.01027.

Key, Brian, Oressia Zalucki, and Deborah J. Brown. 2021. "Neural Design Principles for Subjective Experience: Implications for Insects." *Frontiers in Behavioral Neuroscience* 15. https://doi.org/10.3389/fnbeh.2021.658037.

Key, Brian, Oressia Zalucki, and Deborah J. Brown. 2022. "A First Principles Approach to Subjective Experience." *Frontiers in Systems Neuroscience* 16: 756224. https://doi.org/10.3389/fnsys.2022.756224.

Kolodny, Oren, Roy Moyal, and Shimon Edelman. 2021. "A Possible Evolutionary Function of Phenomenal Conscious Experience of Pain." *Neuroscience of Consciousness* 2021 (2): p.niab012. https://doi.org/10.1093/nc/niab012.

Koppel, Lina, Giovanni Novembre, Robin Kämpe, Mattias Savallampi, India Morrison. 2023. "Prediction and Action in Cortical Pain Processing." *Cerebral Cortex* 33 (3): 794–810. https://doi.org/10.1093/cercor/bhac102.

Lavaud, Simon, Mattia D'Andola, Charlotte Bichara, and Aya Takeoka. 2022. "Electrophysiological Signatures Reveal Spinal Learning Mechanisms for a Lasting Sensorimotor Adaptation." *bioRxiv*. https://doi.org/10.1101/2022.03.30.486422.

Le Bars, Daniel, Manuela Gozariu, and Samuel W. Cadden. 2001. "Animal Models of Nociception." *Pharmacological Reviews* 53 (4): 597–652.

Lewis, David. 1980. "Mad Pain and Martian Pain." In *The Language and Thought Series*, edited by Ned Block, 216–22. Cambridge, MA and London: Harvard University Press.

Li, Feng, Jack W Lindsey, Elizabeth C Marin, Nils Otto, Marisa Dreher, Georgia Dempsey, Ildiko Stark, et al. 2020. "The Connectome of the Adult Drosophila Mushroom Body Provides Insights into Function." *Elife* 9: e62576. https://doi.org/10.7554/eLife.62576.

Maher, Chauncey. 2020. "Experiment Rather Than Define." *Trends in Plant Science* 25: 213–14. https://doi.org/10.1016/j.tplants.2019.12.014.

Mancini, Flavia, Suyi Zhang, and Ben Seymour. 2022. "Computational and Neural Mechanisms of Statistical Pain Learning." *Nature Communications* 13: 6613. https://doi.org/10.1038/s41467-022-34283-9.

Marder, Michael. 2012. "Plant Intentionality and the Phenomenological Framework of Plant Intelligence." *Plant Signaling & Behavior* 7: 1365–72. https://doi.org/10.4161/psb.21954.

Marder, Michael. 2014. *The Philosopher's Plant: An Intellectual Herbarium.* New York: Columbia University Press.

Nagasako, Elna M., Anne Louise Oaklander, Robert H. Dworkin. 2003. "Congenital Insensitivity to Pain: An Update." *Pain*, 101 (3): 213–19. https://doi.org/10.1016/s0304-3959(02)00482-7.

Nagashima, Ayumi., Takumi Higaki, Takao Koeduka, Ken Ishigami, Satoko Hosokawa, Hidenori Watanabe, Kenji Matsui, et al. 2019. "Transcriptional Regulators Involved in Responses to Volatile Organic Compounds in Plants." *Journal of Biological Chemistry* 294 (7): 2256–66. https://doi.org/10.1074/jbc.ra118.005843.

Nash, Tim P. 2005. "Editorial II: What Use is Pain?" *British Journal of Anaesthesia* 94 (2): 146–9. https://doi.org/10.1093/bja/aei017.

Owald, D., Johannes Felsenberg, Clifford B. Talbot, Gaurav Das, Emmanuel Perisse, Wolf Huetteroth, and Scott Waddell. 2015. "Activity of Defined Mushroom Body Output Neurons Underlies Learned Olfactory Behavior in Drosophila." *Neuron* 86 (2): 417–27. https://doi.org/10.1016/j.neuron.2015.03.025.

Parise, André G., Monica Gagliano, and Gustavo M. Souza. 2020. "Extended Cognition in Plants: Is It Possible?" *Plant Signaling & Behavior* 15 (2): 1710661. https://doi.org/10.1080/15592324.2019.1710661.

Pelizzon, Alessandro, and Monica Gagliano. 2015. "The Sentience of Plants: Animal Rights and Rights of Nature Intersecting." *Australia Animal Protection Law Journal* 11: 5.

Perisse, Emmanuel, Magdalena Miranda, and Stéphanie Trouche. 2023. "Modulation of Aversive Value Coding in the Vertebrate and Invertebrate Brain." *Current Opinion in Neurobiology* 79: 102696. https://doi.org/10.1016/j.conb.2023.102696.

Peterson, Carrie L., Zachary A. Riley, Eileen T. Krepkovich, Wendy M. Murray, and Eric J. Perreault. 2014. "Withdrawal Reflexes in the Upper Limb Adapt to Arm Posture and Stimulus Location." *Muscle & Nerve* 49 (5): 716–23. https://doi.org/10.1002/mus.23987.

Puce, Alina, Truett Allison, John C. Gore, and Gregory McCarthy. 1995. "Face-sensitive Regions in Human Extrastriate Cortex Studied by Functional MRI." *Journal of Neurophysiology* 74: 1192–9. https://doi.org/10.1152/jn.1995.74.3.1192.

Riemensperger, Thomas, Thomas Völler, Patrick Stock, Erich Buchner, and André Fiala. 2005. "Punishment Prediction by Dopaminergic Neurons in Drosophila." *Current Biology* 15 (21): 1953–60. https://doi.org/10.1016/j.cub.2005.09.042.

Robinson, David G., Andreas Draguhn, and Lincoln Taiz. 2020. "Plant 'Intelligence' Changes Nothing." *EMBO Reports* 21 (5): e50395. https://doi.org/10.15252/embr.202050395.

Sahraie, A., L. Weiskrantz, J. L. Barbur, A. Simmons, S. C. R. Williams, and M. J. Brammer. 1997. "Pattern of Neuronal Activity Associated with Conscious and Unconscious Processing of Visual Signals." *Proceedings of the National Academy of Sciences* 94 (17): 9406–11. https://doi.org/10.1073/pnas.94.17.9406.

Segundo-Ortin, Miguel, and Paco Calvo. 2023. "Plant Sentience? Between Romanticism and Denial: Science." *Animal Sentience* 33 (1). http://dx.doi.org/10.51291/2377-7478.1772.

Sherrington, C. S. 1910. "Flexion-reflex of the Limb, Crossed Extension-reflex, and Reflex Stepping and Standing." *Journal of Physiology* 40 (1–2): 28–121. https://doi.org/10.1113/jphysiol.1910.sp001362.

Shurrager, P. S., and Elmer Culler. 1940. "Conditioning in the Spinal Dog." *Journal of Experimental Psychology* 26 (2): 133. https://psycnet.apa.org/doi/10.1037/h0054950.

Siju, K. P., Jean Francois De Backer, and Ilona C. Grunwald Kadow. 2021. "Dopamine Modulation of Sensory Processing and Adaptive Behavior in Flies." *Cell and Tissue Research* 383: 207–25. https://doi.org/10.1007/s00441-020-03371-x.

Struik, Paul C. 2023. "Plants Detect and Adapt, But do not Feel." *Animal Sentience* 8 (33): 3. http://dx.doi.org/10.51291/2377-7478.1777.

Sun, Guanghao, Michael McCartin, Weizhuo Liu, Qiaosheng Zhang, Zhe Sage Chen, and Jing Wang. 2023. "Temporal Pain Processing in the Primary Somatosensory Cortex and Anterior Cingulate Cortex." *Molecular Brain* 16: 1–8. https://doi.org/10.1186/s13041-022-00991-y.

Trewavas, Anthony. 2017. "The Foundations of Plant Intelligence." *Interface Focus* 7 (3): 20160098. https://doi.org/10.1098/rsfs.2016.0098.

Van Gelder, Tim. 1995. "What Might Cognition be, if not Computation?" *The Journal of Philosophy* 92 (7): 345–81. https://doi.org/10.2307/2941061.

Weisman, Asaf, John Quintner, and Youssef Masharawi. 2019. "Congenital Insensitivity to Pain: A Misnomer." *The Journal of Pain* 20 (9): 1011–14. https://doi.org/10.1016/j.jpain.2019.01.331.

Wong, P. S., E. Bernat, S. Bunce, and H. Shevrin. 1997. "Brain Indices of Nonconscious Associative Learning." *Consciousness and Cognition* 6 (4): 519–44. https://doi.org/10.1006/ccog.1997.0322.

Wu, Jianqang. 2023. "Plant Biology: Young Maize Leaves 'Smell' a Volatile Danger Signal." *Current Biology* 33: R914–16. https://doi.org/10.1016/j.cub.2023.07.034.

Yamada, D., Daniel Bushey, Feng Li, Karen L. Hibbard, Megan Sammons, Jan Funke, Ashok Litwin-Kumar, et al. 2023. "Hierarchical Architecture of Dopaminergic Circuits Enables Second-order Conditioning in Drosophila." *Elife* 12: e79042. https://doi.org/10.7554/elife.79042.

Yang, Tsu-Hat., Aurore Chetelat, Andrzej Kurenda, and Edward E. Farmer. 2023. "Mechanosensation in Leaf Veins." *Science Advances* 9 (38): eadh5078. https://doi.org/10.1126/sciadv.adh5078.

Yokawa, Ken, and František Baluška. 2018. "Fish and Plant Sentience: Anesthetized Plants and Fishes Cannot Respond to Stimuli." *Animal Sentience* 21 (6). http://dx.doi.org/10.51291/2377-7478.1329.

Zalucki, Oressia, Deborah J. Brown, and Brian Key. 2023. "What if Worms Were Sentient? Insights into Subjective Experience from the Caenorhabditis Elegans Connectome." *Biology & Philosophy* 38 (5): 34. https://doi.org/10.1007/s10539-023-09924-y.

10 A Liberal View on Plant Consciousness

Markus Wild

10.1 The Concept of Consciousness: What Is It Like?

This chapter is about the contested question of plant consciousness. The concept of CONSCIOUSNESS I'm interested in refers to phenomenal consciousness.[1] According to standard use of the term 'consciousness' in the philosophy of mind and consciousness science, being conscious is the subjective experiential feeling of being in a certain state. Though a rock or a bicycle can be cold or kicked around, it is (most people would probably say) nothing like to be a cold rock or a kicked bicycle. The same is not true in your case or in the case of a pig (most people would say). It feels somehow to you or to the pig to be cold or to be kicked. However, there's nothing it is like to be a rock or a bicycle.[2] CONSCIOUSNESS in this sense can be defined in the following way:

> A being is phenomenally conscious if and only if it has some phenomenally conscious states. A state is phenomenally conscious if and only if there is something it is like to be in that state (Nagel 1974). In other words, phenomenally conscious states are felt from a first-person point of view which is to say that they are experienced subjectively.
> (Dung and Newen 2023)

For a conscious creature there is something it is like to be that creature (it has subjective experience), and for a conscious state there is something it is like for that creature to be in that state (they are experienced subjectively). Call this the 'Standard Definition' (SD) and call the relevant feature of SD 'Subjectivity'. As the quote illustrates, it is part of SD to define creature consciousness through state consciousness. Call this feature 'Derivativeness'. Moreover, the property of consciousness is a categorical or determinate property, since consciousness is, as many philosophers claim, an all or nothing affair. Call this 'Determinacy'. Note that Determinacy is sometimes intended to deny degrees of consciousness and sometimes borderline cases of consciousness. While I think that Subjectivity is the core feature

DOI: 10.4324/9781003393375-15

of SD, I think that both Derivativeness and Determinacy are improper and should not be part of SD. I'll come back to this crucial point in the last section of this chapter.

Note that SD does not imply any particular *theory* of consciousness, nor presuppose any specific *methods* for determining whether a creature or state are conscious. Sometimes, for example, consciousness is *defined* as the ability to integrate information, thus confusing explanans and explanandum. However, the Information Integration Theory is intended to *explain* consciousness and not to *define* it (Tononi and Koch 2015). Consciousness is also sometimes defined in terms of the *methods* used to infer whether a creature or a state is conscious, confusing, as it were, premises and conclusion. For example, those who would *define* consciousness in terms of certain learning capacities are guilty of this confusion. A certain capacity or behaviour can certainly be an indicator of consciousness, but it is neither identical to consciousness nor does it define the concept.

Another point of caution. Despite the fact that some theories identify consciousness with representations of a certain kind (First Order Representational Theories) or with a (unconscious) mental state becoming the target of a higher order representation (Higher Order Representational Theories), SD does not by itself imply representational content. Both defenders of FOR Theories (Tye 1995; Dretske 1995) and of HOR Theories (Lycan 1996; Rosenthal 1997; Carruthers 2000) ground consciousness in representational content, while another group of philosophers grounds the essential feature of representations, namely intentionality, in consciousness (Searle 1990; Siewer 1998; Horgan and Tienson 2002). While both groups agree that consciousness in the relevant sense is defined by SD, they disagree about the constitutive relation between consciousness and representational contents. Thus, a creature can have sensations without representing objects distinct from that creature itself. While perceptions can be understood in this way (if a perception is a mental representation of some object distinct from the perceiving subject), and while states of perceptions certainly can become conscious states (it feels something like to hear a bird or to smell a fruit), it's not necessary that only representations can be conscious. A sensation (a state of sensing without representational awareness of an object distinct from the subject of awareness itself) could, given certain conditions, feel like something to the subject.

While participants in the debate on plant consciousness use different terms to refer to the consciousness of plants, they share SD. Some authors speak of 'sentience' or 'feeling' instead of 'consciousness'. For example, defenders of plant consciousness define 'sentience' as "the capacity to feel" (Calvo, Sahi, and Trewavas 2017, 2859), as the "capacity of individuals to have felt states" (Segundo-Ortin and Calvo 2023), or say that "consciousness is marked by the presence of phenomenal experience of some

212 of M (page 222)

sort" (Segundo-Ortin and Calvo 2022, 2). Opponents of plant consciousness also accept SD. Here are two examples from two important papers: "Consciousness is defined as the capacity of an organism to have experiences, to feel sensations" (Draghun et al. 2021, 239); "Primary consciousness: sensory consciousness, the basic ability to have subjective experiences, 'something it is like to be'" (Taiz et al. 2019, 680). A minority of authors in the debate draw distinctions between 'consciousness' and 'sentience' (Nani, Volpara, and Faggio 2021) or between 'consciousness' and 'experience' and 'sentience' (Frazier 2021). However, for the sake of clarity I will not care for any differences between these ways of speaking. While Nani, Volpara and Faggio (2021) and Frazier (2021) use 'sentience' in the way I use 'consciousness', they use 'consciousness' not in the way I intend to use it. For the purpose of this chapter, I consider 'consciousness', 'sentience', 'experience' and 'feeling' to be synonyms, and I'll continue using the term 'consciousness' in the sense just defined. In other words, expressions like 'consciousness', 'sentience', 'experience' and 'feeling' express the same concept, namely, CONSCIOUSNESS.

Unfortunately, not everyone contributing to the literature on plant consciousness uses the concept in the same way. An important conceptual confusion is between CONSCIOUSNESS and the concept of awareness. Some papers define 'consciousness' or 'sentience' as "awareness of the external world" (Calvo, Frantisek, and Anthony 2021, 159) or "the capacity to be aware of the environment and to integrate sensory information to purposeful organism's behavior" (Trewavas et al. 2020, 216). Sometimes 'consciousness' or 'sentience' is defined in a hybrid way as "feelings, subjective states, a primitive awareness of events, including awareness of internal states" (Baluška and Reber 2019, 2). However, awareness of the external world (the environment) or of some internal state is not necessarily linked to consciousness. A creature can be aware of something (external or internal) without being in a conscious state, and a creature can be in a conscious state without being aware of something (external or internal). Awareness of the external world or internal states can be a good indicator of the presence of consciousness, but (transitive) awareness is not identical to (intransitive) consciousness. Consider the question of fish pain (Braithwaite 2010; Sneddon 2015; Woodruff 2017; but see Key 2015). Pain is a conscious state. For the fish (if it feels pain), it somehow feels like being in pain. How do we infer that the fish feels pain? Apart from the physiological conditions (the nociceptive system), we notice that the fish tries to nurse the injured area, that it behaves in an erratic manner, that it does not follow its normal behavioural routines, that it tries to get rid of something, for example by going to a part of the aquarium that the resourceful experimenter has enriched with painkillers. The nursing behaviour directed at a specific area of its body shows that the fish is aware of the injured area, but without the additional indicators, this awareness alone does not show that the fish is experiencing pain. The animal's awareness of the injured area is part of the premises that allow the inference to its feeling pain. In general, a living being can be aware of an object or a state without the living being having a subjective experience, in other words, the state or the object need not play a role in the phenomenal experience (cf. Ferretti 2021, 3320–24).

Here is what will happen in this chapter: In the following Section 2, I will present a brief analysis of the plant consciousness debate. In Section 3, I will take a sideways look at the debate on plant cognition and argue that this debate should serve as a model for the debate on plant consciousness. The key point is that the plant cognition debate is not only about empirical evidence in favour of or against plant cognition, but also about different concepts of COGNITION. Accordingly, in Section 4 I turn to two attempts to interpret the use of the predicate 'consciousness' with regard to plants that I consider uninformative or over-restrictive. In Section 5, I adopt a liberal view of the phenomenon of consciousness, rejecting both Determinacy and Derivativeness. I argue that consciousness in plants could be determinate or indeterminate, dichotomous or degreed, state consciousness or non-derivative creature consciousness. This liberal view of consciousness, I hope, opens up more conceptual space in the debate.

10.2 The Debate on Plant Consciousness: Where Do We Stand?

Are plants conscious? Is it something like to be a daffodil? The Romantic poet William Wordsworth seemed to be convinced that it is something like to be a daffodil. In his famous poem "I wandered lonely as a cloud" (1815) he writes that he encountered "A host of golden daffodils […] Fluttering and dancing in the breeze". As he saw it, the daffodils were moving "in glee". With a considerable amount of poetic licence, Wordsworth calls the consciousness of the flowers their "glee". So, gleefulness is what it is like for the daffodils to be in this situation, namely, fluttering and dancing in the wind.

The ongoing debate over plant consciousness wants to move this sort of question from romanticism into the realm of science. It is fair to say that about three decades ago, research into plant consciousness was a hobbyhorse for romantics, esoterics and New Agers (Nagel 1997). This has changed in the past decade and we are now in a position to raise meaningful questions about plant consciousness. This certainly feels like progress. The debate about plant consciousness is in some ways comparable to the debate about animal consciousness in the 1960s. With the emergence of cognitive ethology, the long untouchable question of animal consciousness was raised again. Thomas Nagel came up with his now famous definition of consciousness with an eye on the work on orientation and echolocation in bats (Nagel 1974). Since then, numerous markers, indicators and theories of animal consciousness have been developed (Seth, Baars, and Edelman 2005; Sneddon et al. 2014; Birch, Schnell, and Clayton 2020; Dung 2022; Crump et al. 2022; Dung and Newen 2023; Veit 2023). As a result of this development, the position that vertebrates enjoy or suffer conscious experiences is no longer a romantic musing or revolutionary science but part of normal science. It could be argued that the study of plant consciousness today is in the same situation as the study of animal consciousness was in the 1960s. By analogy with the progress made in the study of animal consciousness, one could surmise that the more we learn about the cognitive or cognitive-like abilities of plants, the more likely we are to ascribe consciousness to plants; and the more this willingness grows, the more we will apply the types of evidence we apply to animals to plants.[3] In the best case, we will develop procedures that are specific to plants. This speculation about the parallel development of the study of animal consciousness and plant consciousness is not based on the assumption that plants have consciousness but expresses a hypothesis.

Not everyone shares Wordsworth's romantic conviction, of course. A substantial group of critics denies that plants can have consciousness or conscious states (Taiz et al. 2019; Brown and Key 2021; Ginsburg and Jablonka 2021; Mallat et al. 2021; Robinson et al. 2021, 2023; see also the contribution of Brown and Key in this volume). In the eyes of these critics, there are three general reasons that speak against plant consciousness: (i) Plants do not meet the necessary requirements for having consciousness. (ii) Theories of consciousness that allow us to attribute consciousness to plants have significant shortcomings. (iii) Proposals to draw analogies between the properties of plants and conscious animals or humans prove misleading. Let me illustrate these three points of criticism by providing a few examples.

(i) According to Ginsburg and Jablonka (2019, 2021) Unlimited Associative Learning (UAL) is the evolutionary transition marker for consciousness. The conditions for UAL include the following abilities: perception unification and differentiation, temporal depth, global accessibility, goal-directed behaviour, selective attention, intentionality, adaptability and self-other distinction. There is no empirical evidence that any plant species possesses these abilities. Consequently, we cannot attribute consciousness to plants.

(ii) Reber (2018) and Baluška and Reber (2019) have proposed a biomolecular basis for plant consciousness, called "Cellular Basis of Consciousness". According to them, consciousness is an inherent feature of life produced by certain biochemical processes in cells. Plants have the necessary biochemical prerequisites for consciousness. This theory is an example of a biopsychism, the position that consciousness is a vital activity of all living beings (Thompson 2022). If all living beings are conscious and if plants are living beings, then of course plants are conscious. Unfortunately, the argument that consciousness must be an inherent property of life is rather weak. It essentially states that evolution is continuous and that therefore consciousness could not have suddenly appeared. The argument overlooks the fact that new and significant forms and functions can arise in the course of evolution through genetic modification and that a distinction should therefore be made in evolution between continuous and discontinuous variations. Moreover, this approach to consciousness faces an objection similar to the hardest problem facing panpsychism, namely, the combination problem (Chalmers 2016). If cells have their own basic forms of consciousness, it is difficult to see how these entities come together to form conscious creatures like ravens or dogs (Ginsburg and Jablonka 2021, 21).

(iii) The third criticism is the most substantial. Defenders of plant consciousness have proposed various functional analogies between physiological features of animals and plants. For example, it has been argued relying on the idea of Plant Neurobiology (Brenner et al. 2006; Calvo 2016) that plants have some of the same molecules that act as neurotransmitters in animals such as glutamate and gamma-aminobutyric acid (GABA), that both plants and animals have action potentials and complex ways of electrical signalling through the plant vascular system that is similar to the animal nervous system (Lee and Calvo 2023; Segundo-Ortin and Calvo 2023). Critics reply that glutamate and GABA do not function as neurotransmitters in plants, but rather as hormone-like signals involved in growth, development, defence and stress management (Robinson and Draguhn 2021). Since glutamate and GABA, including their receptors, differ from neurotransmitter systems in many ways, the analogy is not evidence for similarities between plant and animal neurobiology. The critics also add that action potentials in plants are utterly different from those in animals: they differ in electrochemical origin, speed, and function and do not imply higher sensory functions (Kjelchova et al. 2021). Importantly, critics claim that there are no backward signals in the electrical communication in plants which contrast sharply with the reciprocal signalling in the brains of mammals (Lamme 2006). The basic argument of the defenders is that plants lack none of the functional structures that are allegedly needed for animals to be conscious; therefore, there is no good reason to exclude the possibility that plants have developed specific structures that underlie their own consciousness (Calvo 2007). The critics emphasise the differences between these structures in plants and animals and go on to point out that there are similarities between structures in plants and animals but that these are irrelevant as functional correlates of consciousness (Robinson et al. 2023, 5).

In addition, defenders refer to behavioural criteria for ascribing consciousness to plants. To find out what animal species do have consciousness, we have to make inferences from animal behaviour and from the cognitive capacities this behaviour expresses. Those inferences aren't fool-proof but justified by an inference to the best explanation. Over the last decade comprehensive overviews have been produced with the goal of classifying indicators of animal consciousness (Seth, Baars, and Edelman 2005; Sneddon et al. 2014; Birch, Schnell, and Clayton 2020; Dung 2022; Crump et al. 2022; Dung and Newen 2023). For example, Birch, Schnell and Clayton (2020) have argued that different forms of learning should be used as indicators for animal consciousness. Recent experimental research suggests that plants are capable of the variety of associative learning called classical conditioning (Gagliano et al. 2016). Even if this evidence were uncontested, it would not constitute a marker for consciousness in plants because according to the animal consciousness literature some forms of learning like classical conditions do not require consciousness experience (cf. Clark and Squire 1998; Greenwald and De Houwer 2017). In contrast to classical conditioning trace conditioning separates the unconditioned and the conditioned stimuli by a temporal gap (Droege et al. 2021). In a recent article Segundo-Ortin and Calvo (2023) cite evidence that "plants can engage in complex decision-making, integrating and weighting information from different parameters and trade-offs, and prioritizing responses to improve the chance of survival". The literature on animal consciousness has embraced motivational trade-off behaviour among the reliable indicators of consciousness. However, as Dung (2023) has pointed out, some authors take the display of motivational trade-offs in plants as a reason to exclude this behaviour from the set of markers of animal consciousness (cf. Mason and Lavery 2022).

So, where do we stand? The last observation is a nice entry point for taking stock. In the context of the debate on plant consciousness, it has to remain an open question whether plants have consciousness or not. The *premise* 'plants are not conscious' cannot play a role in this debate, in contrast to the *conclusion* of the same content, because the question is whether plants *are* conscious or not. For this reason, it would be question begging to insist that an indicator or a theory of consciousness is invalid because it allows us to attribute consciousness to plants. For example, some theorists claim that the Integrated Information Theory (IIT) supports the hypothesis that plants are conscious, and therefore IIT is pseudo-science (IIT-Concerned et al. 2023). While IIT is highly debated, in this context IIT cannot be dismissed simply because it allows for plant consciousness. A closer look at the debate on plant cognition can help to deepen this point.

10.3 The Debate on Plant Cognition: What Can We Learn from It?

A growing body of evidence shows that plants possess abilities associated with cognition, e.g. decision-making, anticipation and learning, but the cognitive status of plants continues to be contested. Contesters sometimes apply a particular definition of cognition to the alleged evidence for plant cognition and conclude that the claim that plants are cognitive beings must be either metaphorical or false (Adams 2018). An example of this is the 'Representation Demarcation Challenge' (Lee 2023). Plants do not fulfil seemingly defining aspects of cognition (e.g. in the classical approach cognition is computation over representation with non-derived content), therefore plants are not cognitive creatures. An important defence of the classical approach is that it allows cognition to be clearly distinguished from behaviour and to be understood as a cause of behaviour. However, a distinction is made in cognitive ethology between associative, representational and intentional explanations of animal behaviour, with all these explanations being cognitive explanations of behaviour (cf. Shettleworth 2010). Even in this discipline, cognition cannot be strictly defined by representation. So why should a fixed definition across all disciplines dedicated to the study of cognition be considered so important?

Defenders of plant cognition sometimes evade representationalism by bringing in non-representationalist theories of cognition such as enactivism or ecological psychology. While some commentators on the debate fear that this kind of dialectic will result in "perennial disagreements over the best way to conceptualise the very nature of cognition" (Lee 2023, 17) others claim that the debate about plant cognition is really a debate about two interrelated issues, namely the abilities of plants *and* the concept COGNITION (Colaço 2022). According to this view, the debate is not about whether plants meet well-defined and agreed-upon criteria for cognition, the relevant concept under scrutiny by the defenders of plant cognition is not an *expression* of what cognition is, but rather a *hypothesis* about what cognition might be. In order to test this hypothesis researchers compare plant abilities to paradigmatic cognitive abilities in humans and other animals in order to highlight similarities and differences between these abilities (Colaço 2022). Thus, researchers proposing alternative accounts of cognition in comparison with the classical account actually *combine* their discussion of empirical findings with novel approaches to cognition. David Colaço (2022) refers to three such approaches, namely the 'biogenetic approach' (Lyons 2006), 'basal cognition' (Levin 2021) and 'cobolism' (Keijzer 2021) combining empirical findings about plants (and other organisms) with alternative conceptions of cognition. Participants in this debate do not share a common notion of what cognition is. Is this a problem? I don't think so. The way we understand COGNITION has considerably changed since the cognitive revolution (Akagi 2018) and we don't need a clear-cut definition of COGNITION in order to make progress (Allen 2017). To hypothesise about cognition is "to put forward a concept of cognition as a conjecture that demarcates a set of phenomena against which it can be tested" (Colaço 2022, 453). Thus, to collect a growing body of evidence for decision-making, anticipation and learning in plants is to look for similarities and differences between paradigm cases of cognition and plant abilities.

Importantly, despite the fact that there are different concepts of COGNITION with different contents and extensions in play, research can examine the degree to which the extensions of those concepts overlap (Akagi 2018; Colaço 2022). The overlap clarifies points of disagreement, while the function of the disagreement is to uncover novel and non-superficial similarities among the abilities of plants and paradigmatically cognitive creatures. To be sure, the nature of this process leaves open the possibility that research might not uncover any interesting similarities, and we might be forced to conclude that plants aren't cognitive creatures. In the opposite case, robust similarities in function, substrate and aetiology of cognition and cognition-like abilities could be found between the abilities of plants and animals, and the concept of COGNITION could be usefully extended. Either way, the process is scientifically valuable because confirmations, modifications and rejections of hypotheses have scientific value. At best, the research on plant cognition will revolutionise cognitive science, and at worst, the acquired evidence will strengthen the distinctiveness of cognition from other biological capacities. Thus, the debate over plant cognition is also a debate over what cognition is, so opponents of plant cognition "should not assume a particular representational account as the basis for their criticism, unless they want to beg the question" (Colaço 2022, 453). From this point of view, it is wrong to say that it is worthless to compare the signalling behaviour of plants with that of the nervous system (Robinson and Draguhn 2021, 8). We simply don't know at the moment whether the similarities will turn out to be robust or not, and if they turn out not to be robust, the endeavour will not have been worthless either.

The detour has shown that the debate on plant cognition is not just about new empirical evidence but also about concepts. For this reason, the debate cannot be decided on empirical grounds alone; it must also be conducted conceptually. Specifically, it is a matter of introducing new and expanded concepts based on the empirical evidence and using the similarities and differences to behavioural abilities and biological structures to test the overlaps between the different concepts and the viability of the corresponding hypotheses about cognition. As we have seen, in addition to the classical concept of cognition, we find at least three alternative and extended concepts in play. In addition, anti-representationalist approaches such as ecological psychology represent radical alternatives (see the chapters by Frazier and Lee and Ponkshe in this volume). We have, thus, a lot of different concepts in play. This is not a Babylonian confusion of languages but part of the development of cognitive science.

10.4 How the Semantics of Predicates Takes the Wrong Turn

I want to suggest that not only the debate on plant cognition, but also the debate on plant consciousness is not exclusively about empirical research, but also about conceptual issues. Therefore, it seems worthwhile turning our attention to the concept itself. Two recent contributions take a look at the predicates 'is conscious' (Maher 2021) and 'consciousness' (Brown and Key 2021). I think that they are on the right track, but they take a wrong turn.[4]

Maher (2021) claims that most people assume that the predicate 'is conscious' is determinate, while he proposes that it is not. He supports his claim by evidence from the history of the predicate and current disagreements among competent users. The explanation for why the predicate 'is conscious' is indeterminate for plants consist in the alleged fact that we rely on various empirical indicators when we characterise human and nonhuman animals as conscious and that plants differ significantly from animals, while also being substantively similar. Maher claims that the extension of the predicate has so far been tacitly restricted to human (and nonhuman) animals. For this reason, its application to plants is indeterminate, by which he means that plants currently neither are nor aren't in its extension. So, it's an open question whether plants will be in the extension of the predicate. Plants, Maher seems to say, are either conscious or they aren't, but at present we're not sure if and how we should apply the predicate to plants. This seems to me a *non*-informative claim, since it seems to be nothing more than a restatement of the ongoing debate about plant consciousness. The defender of plant consciousness do not claim that plants *are* conscious, they claim that it is possible and worth our scientific attention, but we currently don't know if we really should apply the predicate outside the animal realm.

Brown and Key (2021) are more determinate about strict uses of psychological predicates, because they doubt that such predicates allow application beyond rigid use. In particular, the strict use of 'consciousness' does not allow the predicate to be extended to plants. The authors introduce 'consciousness' as a term "referring to subjective human experience" (161). Therefore, the strict use of the predicate should not be extended to plants (and many animals). In particular, they criticise the ambiguity in the use of the predicate:

Ambiguity is seen to assist with making the argument that fish or insects or molluscs or cephalopods, or whatever species is under consideration, are conscious, without implying either that the kind of consciousness such creatures experience is just like ours or that it depends on the same neural circuitry that accounts for consciousness in humans.

(Brown and Key 2021, 163)

For the authors "it is hard to envisage that the term 'consciousness' could lose its association with subjective experience" (162). I agree that CONSCIOUSNESS must not lose its association with subjective experience, this is the feature of SD I called 'Subjectivity' in the first section. However, having a nonhuman subjective experience is simply not the same as having a *human* experience. For example, to say that it is somehow for a fish to be in pain is to say that the fish is having a subjective experience; it is not to say that the fish is having a subjective *human* experience or that it is like *our* experience or that the same neural connections exist in the fish brains and in *our* brains. Nothing in the semantics of 'consciousness' demands that we refer to *human* experience exclusively. If we were to take this suggestion seriously, we could use the expression to refer solely to our individual subjective experiences. If I say 'My fellow human being Mehmet is in pain', I would have to mean that Mehmet suffers my subjective experience of pain, which is obviously wrong. Presumably, one would allow us to apply the predicate to Mehmet because of certain similarities between us, but then there is nothing to prevent us from extending this procedure and applying it to other cases that are similar to humans. Once we have arrived at mammals, we can take the step to fish. Finally, the question arises as to whether we can take the step into the realm of plants. Of course, in all the cases we don't attribute human experience or my subjective experience to those creatures.

As I said, I think that Maher (2021) and Brown and Key (2021) are on the right track, but they take the wrong turn. I suggest to travel the road not taken. Here it is: Maher (2021) explicitly states that his concern with the indeterminacy of the predicate 'is conscious' is not a concern with vagueness, so he doesn't mean that the property of consciousness allows borderline cases. Brown and Key (2021) explicitly criticise the idea that consciousness comes in degrees. I think, however, that both claims actually do make sense.

10.5 The Road Not Taken: Indeterminate, Degreed, Non-derivative Consciousness

The obvious difference between the debates on plant cognition and plant consciousness is that in the former case we are dealing with different concepts of COGNITION while in the latter case we deal with only one concept of CONSCIOUSNESS, namely the concept I introduced at the beginning of this chapter by SD. The core feature of SD is Subjectivity. However, I think that we can and should drop the additional features Determinacy and Derivativeness. With this move, we can develop a more nuanced understanding of consciousness and expand our view beyond Determinacy and Derivativeness. Let's start with Determinacy.

According to Determinacy consciousness is all-or-nothing, which means that it is something to be like a certain creature, or it is not, or a state is like something to be in, or it is not. There are two issues that must be kept separate, namely the idea that conscious is indeterminate (vague) and allows for borderline cases, and the idea that consciousness comes in degrees. Both issues are often confused and should be kept apart.

> If it can be a matter of degree whether some x is F, then F allows for indeterminacy, if not, then F is always determinate. If F comes in degrees, then F is degreed, if not, then F is dichotomous. [...] Questions of determinacy concern membership within categories. Questions about degrees concern values of magnitudes.
>
> (Lee 2022, 558)

Take the property of being warm. It can be indeterminate whether water is warm or not, for example by being luke-warm. Once we definitely think that the water is warm, the water can be more or less warm, thus the property of being warm is degreed. Determinacy implies that conscious is always determinate; however, it doesn't imply that consciousness is dichotomous. I think that consciousness is both indeterminate and degreed. Thus, I take the rejection of Determinacy to imply that it's a matter of degree whether x is conscious (there are borderline cases) and that consciousness may, in addition, come in degrees. Thus, a creature or a state x can be more conscious than a creature or a state y. Given this point of view, it is possible to claim that plants are borderline conscious while (say) vertebrates are conscious.[5]

Let's start with indeterminacy. Indeterminacy allows for borderline consciousness. A creature is borderline conscious if it enters a borderline conscious state or states but is never determinately conscious. It has been claimed that the indeterminacy of consciousness is a consequence of many of our best theories of consciousness. Here's an argument to this consequence: Consciousness is either determinate or indeterminate. If consciousness is determinate, then it isn't a physical or biological phenomenon. However, according to our best theories of consciousness, consciousness is a physical or biological phenomenon. Therefore, consciousness is indeterminate. To illustrate the main point of this line of reasoning, take the theory that consciousness is one and the same as neuronal oscillation of

40MHz. This should not rule out every neuronal oscillation that is not *exactly* 40MHz, rather consciousness is identical to neuronal oscillation sufficiently close to 40MHz. Since 'sufficiently close' indicates a vague property, consciousness is a vague property. (The example is taken from Tye 2021, 7.) Because the same kind of argument could be repeated for any naturalistic theory of consciousness, according to these theories consciousness is an indeterminate or vague property. Since all participants in the debate about plant conscious seem to accept some naturalistic theory of consciousness, they should allow that consciousness is an indeterminate property.

Why do most philosophers think otherwise? The general gist of the scepticism against vague consciousness goes something like this: If consciousness were indeterminate, we should be able to conceive of borderline cases, but we can't, so consciousness is not indeterminate. The inconceivability of borderline consciousness implies its impossibility. As Schwitzgebel (2023) has recently argued, this argument fails. The conceivability of borderline cases requires conceiving determinedly what it would be like to be in a borderline state, but such states are not determinately like anything, therefore it's not legitimate to ask for conceivable examples of borderline conscious states (cf. Hall 2023). Thus, there could be borderline cases of consciousness, and plants might be just that: borderline conscious creatures. If consciousness is indeterminate it is neither determinately true nor determinately false that a plant-state or a plant is conscious. It would only be true that they would be indeterminately or vaguely conscious. How could we know whether a creature is borderline conscious? Only indirectly. Imagine that consciousness requires that creatures are able to learn in certain ways, but not in the sense of classical conditioning. Imagine further that a first group of plants is capable of classical conditioning but notof other forms of learning, while a second group of plants is not even capable of classical conditioning. The second group would have no consciousness and the first group could be borderline conscious.

Let's turn to degrees. If consciousness is a degreed property it comes in degrees. Examples for degreed properties are mass, size, warmth, or health; examples for dichotomous properties are being a prime number, nationality, or species. I can be more or less healthy, but I can't be more or less Swiss (in the sense of being a Swiss citizen, not of being rich, clean, punctual, or able to yodel). It's important to emphasise, however, that the thesis that a property like consciousness comes in degrees does not mean that an associated property is degreed (Lee 2022, 556). For example, I can be more or less aware of something, or I can be more or less awake. However, being more or less conscious is not the same as being more or less aware or awake. The degrees concern only degrees of consciousness according to SD. Moreover, it is not about certain features of conscious states such as intensity, vividness or richness. Conscious states may be more or less intense, vivid or rich, but the thesis of degreed consciousness implies that the consciousness of a state or a creature is itself degreed. And finally, the thesis does not imply continuity. In the case of a degreed property, the values of this property can be continuous or non-continuous. For instance, x can be more conscious than y without there being continuity between x and y. We must note an important distinction here (Lee 2022, 561–2). According to SD for x to be conscious is for x to be something it is like to be x. Now, it sounds obviously odd to say 'There's something it is like to be x more than it is like to be y'. But note that the property 'is conscious' is distinct from the property 'consciousness'. Even if 'x is conscious' does not come in degrees, it does not follow that 'consciousness' does not come in degrees. If a plant is conscious, then it is something like to be that plant, without degrees. Similarly, an object with a size has a certain size, the object does not have size in degrees, yet this does not imply that consciousness or size are not degreed properties. Thus, object x could be bigger than object y, but the size of x is not degreed. In the same vein, a plant could have less consciousness than another plant or plants could be more conscious than insects or mushrooms, or the plants of a certain species could have more consciousness than those of another species.

The degree thesis allows for differences and similarities. Take whatever your favourite theory of consciousness is. Call the property that explains consciousness P. If the amount of P in x is not the same as the amount of P in y, then what it is like to be x differs from what it is like to be y. If the amount of P in x is bigger than the amount of P in y, and if the amount of P in y is bigger than the amount of P in z, then what it is like the be x is more similar to what it is like to be y than what it is like to be z. Lee (2022, 565) calls this the "Difference Criterion" and the "Similarity Criterion", respectively. By introducing degrees of consciousness, differences and similarities between the consciousness of plants and other living beings can be articulated. This option corresponds to the idea formulated above with reference to the debate on plant cognition. The idea does not consist in the application of a specific concept of COGNITION, it rather formulates a hypothesis with the aim of either expanding or sharpening the concept by comparing paradigmatically cognitive creatures (such as mammals or vertebrates) and plants by identifying similarities and differences. The same can be said about plant consciousness, since the idea that plants are conscious creatures formulates a hypothesis with the same aim.

I will now turn to the second characteristic of consciousness that I do not consider to be constitutive of consciousness, namely Derivativeness. According to Derivativeness a creature is phenomenally conscious if and only if it has some conscious states. Thus, consciousness is basically state consciousness, and only derivatively creature consciousness. Let's call 'total subjective experience' an experience that completely characterises what it is like to be a certain creature at a given time t. The total subjective experience could either be composed from multiple primitive conscious states (Derivativeness) or it could itself be a primitive conscious state not so composed (non-Derivativeness). Thus, we can ascribe a non-derivative total subjective experience to a creature at t without ascribing more basic conscious states to it. Remember the gleeful daffodils "dancing in the breeze". The gleefulness doesn't imply that the daffodils have conscious states of glee which is the reason for their being gleeful. Rather, we can understand Wordsworth as attributing glee to the daffodils as creatures, but not to individual conscious states that make up plant-gleefulness. It is possible for a conscious creature to enjoy consciousness (i.e. to be in a state of non-derivative total subjective experience) without having constitutive conscious states. Thus, creature consciousness and state consciousness are in a sense, independent of each other: creature consciousness could just be the total conscious state of a system. Taking together my denial of both Determinacy and Derivativeness, I submit that both state consciousness and non-derivative creature consciousness can be indeterminate or degreed.

Some authors claim that animals perceive the world by representing objects in their environment (see the contributions of Matthen and Schulte in this volume), while plants regulate their own bodies based on input of their sensors. Calvo and Friston (2017) hypothesised that plants are examples of radical predictive processing, which means that plants are not in the business of representing the external world by representational states, rather plants control their bodies by a dynamic state of anticipation of sensory states, thus responding to sensed environmental contingencies. This total dynamic state of anticipation could well be a non-derivative total subjective experience of the plant without more basic conscious states. While the sensory states register unconsciously environmental contingencies, the plant's state of anticipation at t is the total subjective experience of the creature at t.

One often cited piece of evidence for plant consciousness involves the use of anaesthetics in plants (Yokawa et al. 2018). The Venus flytrap shuts down snap-shutting behaviour and *M. pudica* stops leaf-folding behaviour under general anaesthesia. Since anaesthesia in humans induces loss of consciousness, the same could occur in plants, thus general anaesthesia not only disrupts plant behaviour but plant consciousness. It has been argued against this hypothesis that anaesthesia alters conscious states in humans and in certain animals and, in particular, modifies states of pain. While there is clear evidence that anaesthetic substances affect non-neural physiological processes in plants, plants do not have the molecular or structural mechanisms to produce states of pain. Consequently, it cannot be concluded from the effects of anaesthesia on plants that plants have conscious states such as pain (Draguhn, Mallatt, and Robinson 2021). But the claim that no specific conscious states such as pain experiences can be ascribed to plants only rules out that anaesthesia makes a difference for conscious *states* of plants, but it doesn't rule out that the non-derivative creature consciousness of a plant (it's total subjective experience) could be put to sleep by anaesthesia. Thus, there could be an interesting difference between the consciousness of plants and the consciousness of animals. It could be the case that animals have numerous conscious states (normally associated with the representation of objects in their environment) that together constitute their creature consciousness, whereas plants enjoy a non-derivative state of total subjective experience (perhaps associated with their dynamic state of anticipation) that is not constituted by conscious states. The daffodils are totally gleeful, they're not mereologically gleeful.

10.6 Conclusion

According to my argument in this chapter, there are the following options for the debate on plant consciousness. (1) All participants should focus primarily on the debate about plant cognition and place the question of plant consciousness under a moratorium. The cognition debate should be understood as a hypothesis about the expansion of the concept COGNITION. Only if this expansion is successful, we should ask ourselves whether or not the cognitive abilities of plants allow an attribution of consciousness comparable to that of mammals and other animals. The reason for this is that we must also rely on behavioural markers and indicators when attributing conscious states to vertebrates. (2) The defenders of plant consciousness continue to refer to questionable theories of consciousness, such as Cellular Basis of Consciousness, according to which plants obviously possess consciousness. But this seems an easy success. As a consequence, the debate will primarily centre on the merits of the corresponding theories of consciousness and not on plant consciousness. (3) The debate should extend to non-orthodox forms of consciousness, so that the debate on plant consciousness, like the debate on plant cognition, can combine empirical research with the conceptual expansion of our understanding of the relevant phenomena. In this way, both empirical and conceptual progress, positive or negative, could be achieved. I suggested that CONSCIOUSNESS could be liberally conceptualised as state consciousness and non-derivative creature consciousness. In addition, I suggested that both state consciousness and non-derivative creature consciousness could be conceptualised as determinate or indeterminate and as degreed or dichotomous.

Acknowledgements

I would like to thank the audiences in Basel and Essen for helpful questions, suggestions and criticisms. Special thanks to Gabriele Ferretti and an anonymous reviewer for valuable and constructive comments.

Notes

1 I use CONSCIOUSNESS to refer to the concept, 'consciousness' to refer to the term and consciousness to refer to the phenomenon itself. The same applies to COGNITION, 'cognition' and cognition.
2 My way of introducing these issues seems to exclude panpsychism from the outset. I will not consider this idea, which has been gaining increasing attention in recent philosophical discussions, for the sole reason that it plays no role in the debate on plant consciousness to date. If one assumes some panpsychist view, the problem of consciousness in plants and animals presents itself in a different way, for example, in the shape of the problem of how the consciousness of the parts of an organism can together create the consciousness of the organism. This is known as the 'combination problem' (Chalmers 2016).
3 Note, however, that there are important hurdles in customising animal protocols for plant research (cf. Ponkshe et al. 2023).
4 Note that the argument of Brown and Key in the present volume against plant sentience is a different and quite strong argument. It is not touched by my criticism of their former paper in this chapter.
5 In a recent paper Kristin Andrews (2024) claims that accepting the hypothesis all animals are conscious will promote research leading to a secure theory of consciousness. From this point of view plants could well be border-line cases.

References

Adams, Fred. 2018. "Cognition Wars." *Studies in History and Philosophy of Science Part A* 68: 20–30. https://doi.org/10.1016/j.shpsa.2017.11.007

Akagi, Mikio. 2018. "Rethinking the Problem of Cognition." *Synthese* 195 (8): 3547–70. https://doi.org/10.1007/s11229-017-1383-2

Allen, Colin. 2017. "On (not) Defining Cognition." *Synthese* 194 (3): 1–17. https://link.springer.com/article/10.1007/s11229-017-1454-4.

Andrews, Kristin 2024. *All Animals Are Conscious: Shifting the Null Hypothesis in Consciousness Science.* Mind&Language. https://doi.org/10.1111/mila.12498

Baluška, František, and Arthur Reber. 2019. "Sentience and Consciousness in Single Cells: How the First Minds Emerged in Unicellular Species." *BioEssays* 41 (3): 1800229. https://doi.org/10.1002/bies.201800229.

Birch, Jonathan, Alexandra K. Schnell, and Nicola S. Clayton (2020). "Dimensions of Animal Consciousness." *Trends in Cognitive Sciences* 24 (10): 789–801. https://doi.org/10.1016/j.tics.2020.07.007.

Braithwaite, Victoria A. 2010. *Do Fish Feel Pain?* Oxford: Oxford University Press.

Brenner, Eric D., Rainer Stahlberg, Stefano Mancuso, Jorge Vivanco, František Baluška, and Elizabeth Van Volkenburgh. 2006. "Plant Neurobiology: An Integrated View of Plant Signaling." *Trends in Plant Science* 11 (8): 413–19. https://doi.org/10.1016/j.tplants.2006.06.009.

Brown, Deborah, and Brian Key. 2021. "Plant Sentience, Semantics, and the Emergentist Dilemma." *Journal of Consciousness Studies* 28 (1–2): 155–83.

Calvo, Paco. 2007. "The Quest for Cognition in Plant Neurobiology." *Plant Signaling & Behavior* 2 (4): 208–11. https://doi.org/10.4161/psb.2.4.4470.

Calvo, Paco. 2016. "The Philosophy of Plant Neurobiology: A Manifesto." *Synthese* 193: 1323–43. https://doi.org/10.1007/s11229-016-1040-1.

Calvo, Paco, Frantisek Baluška, and Anthony Trewavas. 2021. "Integrated Information as a Possible Basis For Plant Consciousness." *Biochemical and Biophysical Communication* 564: 158–65.

Calvo, Paco, and Karl Friston. 2017. "Predicting Green: Really Radical (Plant) Predictive Processing." *Journal of the Royal Society Interface* 14 (131): 20170096. https://doi.org/10.1098/rsif.2017.0096.

Calvo, Paco, Vaidurya Pratap Sahi, and Anthony Trewavas. 2017. "Are Plants Sentient?" *Plant, Cell & Environment* 40 (11): 2858–69. https://doi.org/10.1111/pce.13065.

Carruthers, Peter. 2000. *Phenomenal Consciousness: A Naturalistic Theory.* New York: Cambridge University Press.

Chalmers, David. 2016. "The Combination Problem for Panpsychism." In *Panpsychism*, edited by Godehard Brüntrup, and Ludwig Jaskolla, 179–214. New York: Oxford University Press,.

Clark, Robert E., and Larry R. Squire. 1998. "Classical Conditioning and Brain Systems: The Role of Awareness." *Science* 280 (5360): 77–81. https://doi.org/10.1126/science.280.5360.77.

Colaço, David. 2022. "Why Studying Plant Cognition is Valuable, Even if Plants Aren't Cognitive." *Synthese* 200 (453). https://doi.org/10.1007/s11229-022-03869-7.

Crump, Andrew, Heather Browning, Alexandra Schnell, Charlotte Burn, and Jonathan Birch. 2022. "Sentience in Decapod Crustaceans: A General Framework and Review of the Evidence." *Animal Sentience* 7 (32). http://dx.doi.org/10.51291/2377-7478.1691.

Draguhn, Andreas, Jon M. Mallatt, and David G. Robinson. 2021. "Anesthetics and Plants: No Pain, no Brain, and Therefore no Consciousness." *Protoplasma* 258: 239–48. https://doi.org/10.1007/s00709-020-01550-9.

Dretske, Fred. 1995. *Naturalizing the Mind.* Cambridge, MA: MIT Press.

Droege, Paula, Daniel J. Weiss, Natalie Schwob, and Victoria Braithwaite. 2021. "Trace Conditioning as a Test for Animal Consciousness: A New Approach." *Animal Cognition* 24 (6): 1299–304. https://doi.org/10.1007/s10071-021-01522-3.

Dung, Leonard. 2022. "Assessing Tests of Animal Consciousness." *Consciousness and Cognition* 105: 103410. https://doi.org/10.1016/j.concog.2022.103410.

Dung, Leonard. 2023. "From Animal to Plant Sentience: Is there Credible Evidence?" *Animal Sentience* 33 (10) doi:10.51291/2377-7478.1784

Dung, Leonard, and Albert Newen. 2023. "Profiles of Animal Consciousness: A Species-sensitive, Two-tier Account to Quality and Distribution." *Cognition* 235: 105409. https://doi.org/10.1016/j.cognition.2023.105409.

Ferretti, Gabriele. 2021. "Visual Phenomenology Versus Visuomotor Imagery: How Can We Be Aware of Action Properties?" *Synthese* 198: 3309–38. https://doi.org/10.1007/s11229-019-02282-x.

Frazier, Adrian P. 2021. "On the Possibility of Plant Consciousness: A View from Ecointeractivism." *Mind and Matter* 19 (2): 229–59.

Gagliano, M., V. Vladyslav Vyazovskiy, Alexander A. Borbély, Mavra Grimonprez, and Martial Depczynski. 2016. "Learning by Association in Plants." *Scientific Reports* 6 (1): 38427. https://doi.org/10.1038/srep38427.

Ginsburg, Simona, and Eva Jablonka. 2019. *The Evolution of the Sensitive Soul. Learning and the Origins of Consciousness.* Cambridge, MA: MIT Press.

Ginsburg, Simona, and Eva Jablonka. 2021. "Sentience in Plants: A Green Red Herring?" *Journal of Consciousness Studies* 28 (1–2): 17–33.

Greenwald, Anthony G., and Jan De Houwer. 2017. "Unconscious Conditioning: Demonstration of Existence and Difference from Conscious Conditioning." *Journal of Experimental Psychology: General* 146 (12): 1705–21. https://doi.org/10.1037/xge0000371.

Hall, Geoffrey. 2023. "Is Consciousness Vague?" *Australasian Journal of Philosophy* 101 (3): 670–84. https://doi.org/10.1080/00048402.2022.2036207.

Horgan, Terence, and John Tienson. 2002. "The Intentionality of Phenomenology and the Phenomenology of Intentionality." In *Philosophy of Mind: Classical and Contemporary* Readings, edited by David Chalmers, 520–33. New York: Oxford University Press.

IIT-Concerned, Stephen M. Fleming, Chris Frith, Mel Goodale, Hakwan Lau, Joseph E. LeDoux, Alan L. F. Lee, et al. 2023. "The Integrated Information Theory of Consciousness as Pseudoscience." *PsyArXiv.* https://doi.org/10.31234/osf.io/zsr78.

Keijzer, Fred. 2021. "Demarcating Cognition: The Cognitive Life Sciences." *Synthese* 198 (1): 131–57. https://doi.org/10.1007/s11229-020-02797-8.

Key, Brian. 2015. "Fish Do Not Feel Pain and its Implications for Understanding Phenomenal Consciousness." *Biology and Philososophy* 30 (2):149–65. https://doi: 10.1007/s10539-014-9469-4.

Klejchova, Martina, Fernanda A. L. Silva-Alvim, Michael R. Blatt, and Jonas Chaves Alvim. 2021. "Membrane Voltage Serves as a Dynamic Platform for Spatio-temporal Signaling, Physiological, and Developmental Regulation." *Plant Physiology* 185 (4): 1523–41. https://doi.org/10.1093/plphys/kiab032.

Lamme, Victor A. F. 2006. "Towards a True Neural Stance on Consciousness." *Trends in Cognitive Sciences* 10 (11): 494–501. https://doi.org/10.1016/j.tics.2006.09.001.

Lee, Andrew Y. 2022. "Degrees of Consciousness." *Nous* 57 (3): 553–75. https://doi.org/10.1111/nous.12421.

Lee, Johnny. 2023. What is Cognitive About 'Plant Cognition?', *Biology & Philosophy* 38 (18). https://doi.org/10.1007/s10539-023-09907-z.

Lee, Jonny, and Paco Calvo. 2023. "The Potential of Plant Action Potentials." *Synthese* 202 (6): 176. https://doi.org/10.1007/s11229-023-04398-7.

Levin, Michael. 2021. "Life, Death, and the Self: Fundamental Questions of Primitive Cognition Viewed through the Lens of Body Plasticity and Synthetic Organisms." *Biophysical and Biochmemical Research Communications* 564: 114–33. https://doi.org/10.1016/j.bbrc.2020.10.077.

Lycan, William G. 1996. *Consciousness and Experience.* Cambridge, MA: MIT Press.

Lyons, Pamela. 2006. "The Biogenetic Approach to Cognition." *Cognitive Processing* 7 (1): 11–29. https://doi.org/10.1007/s10339-005-0016-8.

Maher, Chauncey. 2021. "The Indeterminacy of Plant Consciousness." *Journal of Consciousness Studies* 28 (1–2): 136–54.

Mallatt, Jon, Michael R. Blatt, Andreas Draghun, David G. Robinson, and Lincoln Taiz. 2021. "Debunking a Myth: Plant Consciousness." *Protoplasma* 258 (3): 459–76. https://doi.org/10.1007/s00709-020-01579-w.

Mason, Georgia J., and J. Michelle Lavery. 2022. "What is it Like to Be a Bass? Red Herrings, Fish Pain and the Study of Animal Sentience." *Frontiers in Veterinary Science* 9. https://doi.org/10.3389/fvets.2022.948567.

Nagel, Thomas. 1974. "What Is It Like to Be a Bat?" *The Philosophical Review* 83 (4): 435–50.

Nagel, Alexandra H.M. 1997. "Are Plants Conscious?" *Journal of Consciousness Studies* 4(3): 215–30.

Nani, A., G. Volpara, and A. Faggio. 2021. "Sentience With or Without Consciousness." *Journal of Consciousness Studies* 28 (1–2): 60–79.

Ponkshe, Aditya, Jacobo Blancas Barroso, Charles I. Abramson, and Paco Calvo. 2023. "A Case Study of Learning in Plants: Lessons Learned from Pea Plants."

Quarterly Journal of Experimental Psychology. https://doi.org/10.1177/17470218231203078

Reber, Arthur S. 2018. *The First Minds: Caterpillars, Karyotes, and Consciousness.* New York: Oxford University Press.

Robinson, David G, Michael R. Blatt, Andreas Draguhn, Lincoln Taiz, and Jon Mallatt. 2023. "Plants Lack the Functional Neurotransmitters and Signaling Pathways Required for Sentience in Animals." *Animal Sentience* 33 (7) http://dx.doi.org/10.51291/2377-7478.1782.

Robinson, David G., and Andreas Draguhn. 2021. "Plants Have Neither Synapses Nor a Nervous System." *Journal of Plant Physiology* 263: 153467. https://doi.org/10.1016/j.jplph.2021.153467.

Rosenthal, David M. 1997. "A Theory of Consciousness." In *The Nature of Consciousness*, edited by Ned Block, Owen J. Flanagan, and Güven Güzeldere, 729–53. Cambridge, MA: MIT Press.

Schwitzgebel, Eric. 2023. "Borderline Consciousness, When it's Neither Determinately True Nor Determinately False That Experience is Present." *Philosophical Studies* 180: 3415–39. https://doi.org/10.1007/s11098-023-02042-1

Searle, John R. 1990. "Consciousness, Explanatory Inversion and Cognitive Science." *Behavioral and Brain Sciences* 13 (4): 585–642. https://psycnet.apa.org/doi/10.1017/S0140525X00080304.

Segundo-Ortin, Miguel, and Paco Calvo. 2022. "Consciousness and Cognition in Plants." *Wiley Interdisciplinary Reviews: Cognitive Science* 13 (2): e1578. https://doi.org/10.1002/wcs.1578.

Segundo-Ortin, Miguel, and Paco Calvo. 2023. "Plant Sentience? Between Romanticism and Denial: Science." *Animal Sentience* 33 (1). http://dx.doi.org/10.51291/2377-7478.1772.

Seth, Anil K., Bernard J. Baars, and David B. Edelman. 2005. "Criteria for Consciousness in Humans and Other Mammals." *Consciousness and Cognition* 14(1): 119–39. https://doi.org/10.1016/j.concog.2004.08.006.

Shettleworth, Sara J. 2010. *Cognition, Evolution, and Behavior.* 2nd ed. Oxford: Oxford University Press.

Siewer, Charles. 1998. *The Significance of Consciousness.* Princeton: Princeton University Press.

Sneddon, Lynne U. 2015. "Pain in Aquatic Animals." *Journal of Experimental Biology* 218: 967–76. https://doi.org/10.1242/jeb.088823

Sneddon, Lynne U., Robert W. Elwood, Shelley A. Adamo, and Matthew C. Leach. 2014. "Defining and Assessing Animal Pain." *Animal Behaviour* 97: 201–12. https://doi.org/10.1016/j.anbehav.2014.09.007.

Taiz, Lincoln, Daniel Alkon, Andreas Draghun, Angus Murphy, Michael Blatt, Chris Hawes, Gerhard Thiel, and David G. Robinson. 2019. "Plants Neither Possess nor Require Consciousness." *Trends in Plant Science* 24 (8): 677–87. https://doi.org/10.1016/j.tplants.2019.05.008.

Thompson, Evan. 2022. "Could All Life Be Sentient?" *Journal of Consciousness Studies* 29 (3–4): 229–65.

Tononi, Giulio, and Christof Koch. 2015. "Consciousness: Here, There and Everywhere?" *Philosophical Transactions of the Royal Society B* 370: 20140167. https://doi.org/10.1098/rstb.2014.0167.

Trewavas, Anthony, František Baluška, Stefano Mancuso, and Paco Calvo. 2020. "Consciousness Facilitates Plant Behavior." *Trends in Plant Science* 25 (3): 216–17. https://doi.org/10.1016/j.tplants.2019.12.015.

Tye, Michael. 1995. *Ten Problems of Consciousness: A Representational Theory of the Phenomenal Mind.* Cambridge, MA: MIT Press.

Tye, Michael. 2021. *Vagueness and the Evolution of Consciousness: Through the Looking Glass.* New York: Oxford University Press.

Veit, Walter. 2023. *A Philosophy for the Science of Animal* Consciousness. New York: Routledge.

Woodruff, Michael L. 2017. "Consciousness in Teleosts: There is Something it Feel Like to be a Fish." *Animal Sentience* 13 (1). https:// 10.51291/2377-7478.1321

Wordsworth, William. 1815. *Poems by William Wordsworth Including Lyrical Ballads and the Miscellaneous Pieces of the Author,* vol. 1. London: Longman, Hurst, Rees, Orme, and Brown.

Yokawa, Ken, Kagenishi, Andrej Pavlovic, S. Gall, M. Weiland, Stefano Mancuso, and František Baluška. 2018. "Anaesthetics Stop Diverse Plant Organ Movements, Affect Endocytic Vesicle Recycling and ROS Homeostasis, and Block Action Potentials in Venus Flytraps." *Annals of Botany* 122 (5): 747–56. https://doi.org/10.1093/aob/mcx155.

Index

Note: Page locators followed by 'n' refer to notes; Locators in italics refer to figures.

For Product Safety Concerns and Information please contact our EU
representative GPSR@taylorandfrancis.com
Taylor & Francis Verlag GmbH, Kaufingerstraße 24, 80331 München, Germany

9 7 8 1 0 3 2 4 9 3 5 2 7